List of Contents

Abstract

There are many reasons why it is timely to review the development of the mammalian kidney. Perhaps the most important of these is the increasing amount of evidence to demonstrate that factors which impinge on/alter the normal developmental processes of this organ can have lifelong consequences for the health of the adult. The 'Developmental Origins of Health and Adult Disease' (DOHaD) hypothesis, proposes that changes in the environment during the development of an organ or system, can have permanent deleterious effects leading to increased risk of cardiovascular and/or metabolic disease. The permanent metanephric kidney has been shown to be very vulnerable to such influences with many factors shown to alter both the permanent structure and the level of expression of important functional genes. Thus it is important to understand the precise timing of kidney development in terms of both structure and the genes involved at each stage. Such knowledge has been gained by significant advances in technology, which allow quantification of the number of nephrons by unbiased stereology, detections of both levels and site of gene expression, 'knock-out' and knock-in' of genes in animal (mainly mouse) models and by the ability to examine nephron development, in real time, in culture systems.

1
Introduction

There are many reasons why it is timely to review the development of the mammalian kidney. Perhaps the most important of these is the increasing amount of evidence demonstrating that factors which impinge on or alter the normal developmental processes of this organ can have lifelong consequences for the health of the adult. The original Barker hypothesis, more recently termed developmental origins of health and adult disease (DOHaD) hypothesis, proposed that changes in the environment (such as level of nutrition [total, protein, mineral, vitamin] or exposure to stress hormones) during the development of an organ or system, could have permanent deleterious effects leading to increased risk of cardiovascular and/or metabolic disease (Barker and Bagby 2005; Barker 2007; Gluckman and Hanson 2006; Hoy et al. 2005; Moritz et al. 2003, 2005a; Moritz and Bertram 2006).

The permanent metanephric kidney has been shown to be very vulnerable to such influences, with many factors shown to be able to alter both the permanent structure and the level of expression of important functional genes, most likely by the process of epigenetics (Moritz et al. 2003; Bagby 2006; Zandi-Nejad et al. 2006). Thus it has become increasingly important to understand the precise timing of kidney development in terms of both structure and the genes involved at each stage. Such knowledge has been gained by significant advances in technology, which allow quantification of the number of branching points and whole nephrons by unbiased stereology, detections of both levels (microarray, real-time PCR) and site (hybridization histochemistry) of gene expression, and by the ability to examine nephron development, in real time, in culture systems (3D, 4D microscopy) (Caruana et al. 2006b; Sanna-Cherchi et al. 2007; Jain et al. 2007; Bertram 1995, 2001). In addition, knowledge of the relative importance of individual genes in kidney development has been gained from knock-out and knock-in of genes in animal (mainly mouse) models.

The purpose of this review is to examine recent progress in the field of renal development and the long-term impact that poor renal development has on adult health.

2
Morphological Development of the Mammalian Kidney

In mammals, three pairs of excretory organs form from the intermediate mesoderm in a cranial to caudal direction. These are the pronephroi, mesonephroi and metanephroi, respectively. The pronephroi and mesonephroi are transient organs, but their existence is required for the development of the metanephroi or permanent kidneys. The development of these three excretory organs is shown diagrammatically in Fig. 1.

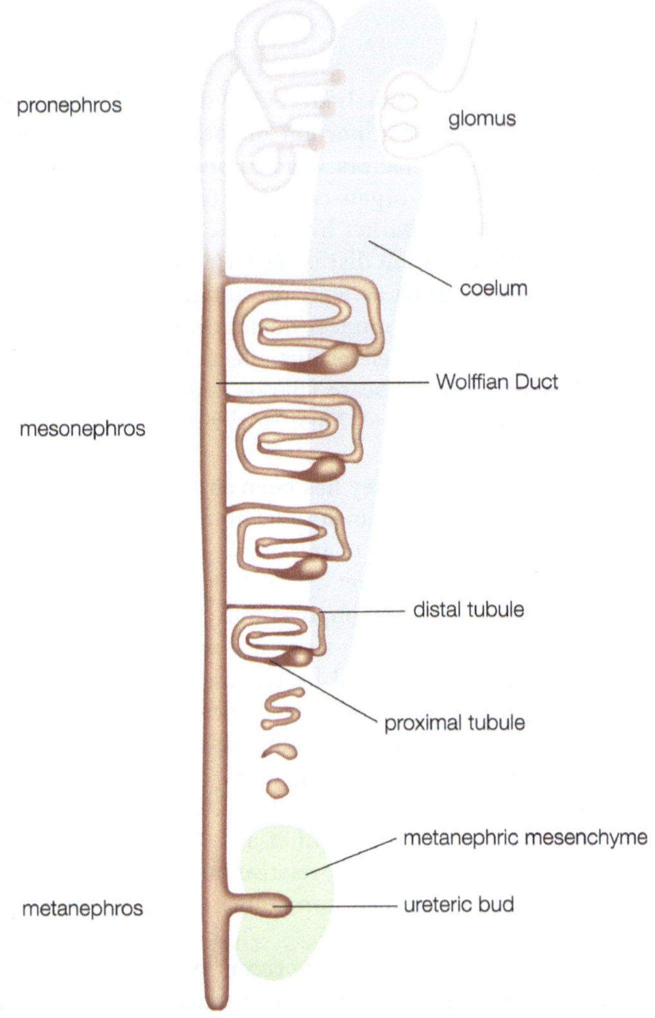

Fig. 1 Schematic representation showing development of the pronephros, mesonephros and metanephros. The pronephros, composed of a single glomus, projects into the nephrocoel but filters directly into the coelom (see text for details) and is depicted as having already degenerated in this schematic diagram. The mesonephros consists of multiple nephrons that develop in a cranial to caudal fashion such that the most caudal structures (renal vesicle, comma and S-shaped bodies) are still in the process of developing into complete nephrons that will attach to the Wolffian duct. The metanephros (11 dpc in the mouse) at this stage comprises the ureteric bud, which has entered the metanephric mesenchyme. (Illustration by Ms. Nina Bosanac, Technology Services Group, Multimedia Services, Monash University)

2.1
Pronephros

The pronephros is the obligatory precursor of the adult renal system. Pronephroi begin to form at 21–22 days post-coitum (dpc) in humans and 8 dpc in the mouse (Woolf et al. 2003). The pronephros develops as a nephrotome containing a nephrocoel (cavity) which is in direct communication with the adjacent intra-embryonic coelom by a short peritoneal funnel. The vascularized filtration unit (glomus) projects into the nephrocoel but filters directly into the coelom. The wall of the nephrotome opposite the peritoneal funnel gives rise to a tubule which joins the nephrocoel to the pronephric duct (future Wolffian duct) via the pronephric tubules. The ciliated cells in the nephrotome move the fluid into pronephric tubules by ciliary action, from where some reabsorption can occur into a surrounding blood sinus. The pronephric duct ends in the cloaca. The number of pronephric tubules varies between different animals and species and in ungulates these tubules are represented by a giant glomerulus in the head of the mesonephros (Vize et al. 1997; reviewed in Vize et al. 2003).

In general, amphibians and fish have well-developed functional pronephroi that persist throughout the life of the organism and regulate water and solute balance as well as blood pH (Drummond and Majumdar 2003). However, amniotes have rudimentary, transient pronephroi which are thought to have no renal function, although this has not been tested. Many of the same genes (*Pax2*, *Pax8*, *Gata3*, *Lim1*, *FGF8*, *Six1*, *WT-1* and *Wnt-4*) are expressed in the pronephros as in the later forms of renal tissue (Vize et al. 1997; Hensey et al. 2002; Chan and Asshima 2006; Bouchard et al. 2002; Grote et al. 2006).

2.2
Mesonephros

As the pronephros regresses, the mesonephros begins to develop as the second excretory organ at 25 dpc in humans (Ludwig and Landmann 2005) and 9 dpc in mice. During early development, the mesonephros occupies a prominent position in the abdominal region of a number of species (Fig. 2). The Wolffian duct (WD) induces the nephrogenic cord or mesonephric mesenchyme to aggregate and undergo a mesenchyme to epithelial transformation (MET) to form a renal vesicle. These inductive signals are similar to those used to induce the metanephric mesenchyme to undergo MET to form a renal vesicle (see Sect. 2.3). The renal vesicle differentiates into an S-shaped structure that elongates and eventually forms a proximal tubule. However, no loop of Henle or juxtaglomerular apparatus form as these tubules connect to the WD. The mesonephroi, with simple but complete nephrons, are the first functioning excretory units in mammals producing small amounts of urine (see Sect. 6.1). These mesonephric structures are transient, with a maximum number of up to 40 mesonephric tubules present at any one time in humans (Sainio and Raatikainen-Ahokas 1999; Ludwig and Landmann 2005). Since the mesonephros begins

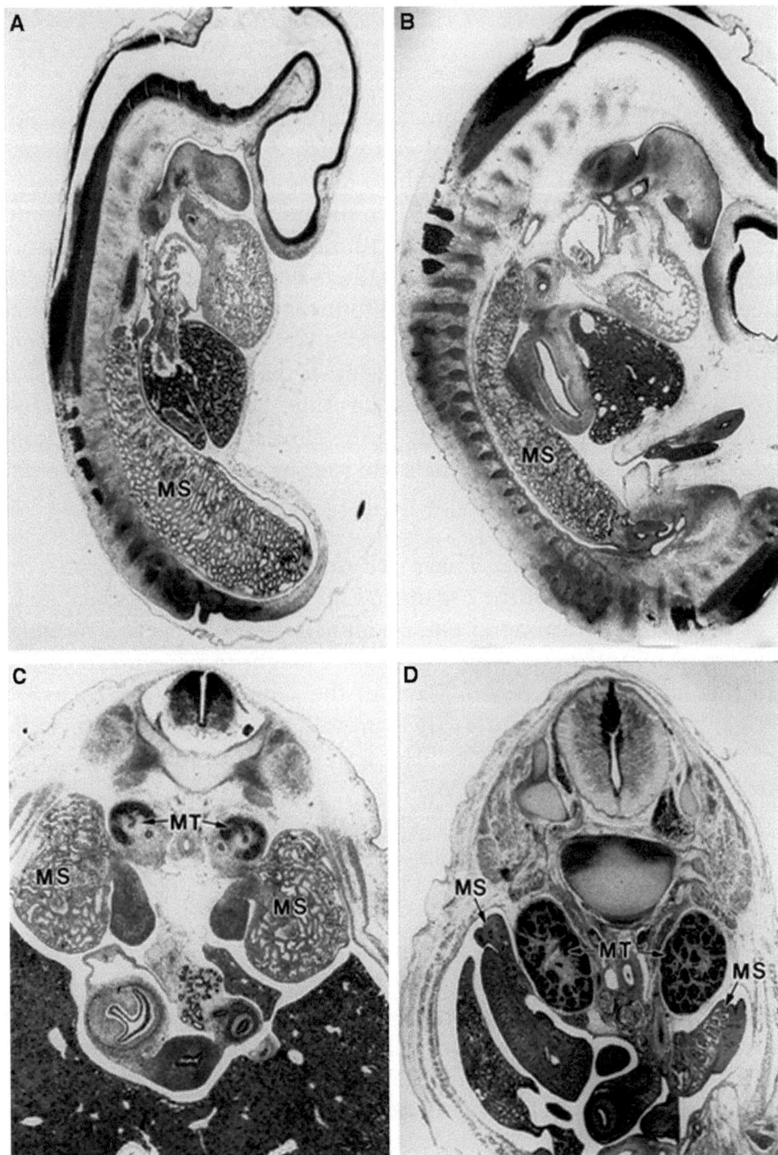

Fig. 2A–D Sagittal (**A** and **B**) and transverse (**C** and **D**) paraffin sections of whole embryos demonstrating the development of the mesonephros. **A** In the pig embryo at 8 mm and **B** the sheep embryo at 27 days of gestation, the mesonephros is the most prominent organ in the future abdominal region. At later stages of development, both the mesonephros and the metanephros can be identified (**C**) in the sheep embryo at 40 days of gestation and (**D**) in the human fetus at 8 weeks of gestation. **C** and **D** show the mesonephros has begun to regress and in the human embryo is less prominent in the abdominal cavity. *MS* mesonephron, *MT* metanephros. (Reproduced with permission from Wintour, Alcorn and Rockell)

its development at the cranial segment of the intermediate mesoderm, the cranial nephrons are more developed and subsequently atrophy and degenerate first in accordance with the cranio-caudal developmental wave in humans (Pole et al. 2002; Carev et al. 2006; Sainio 2003). By 33 dpc in humans, there are characteristic fully differentiated nephrons with the maximum number of nephrons present in the mesonephros at this time point. At this stage, the caudal region of the WD gives rise to the ureteric bud (UB) and the development of the metanephros begins. In females, the mesonephros fully regresses, but in males, remaining mesonephric tubules form the efferent tubules of the testis. As with the pronephros, many of the genes (*Pax2, Wt-1, Osr1, Wnt9b, Six1*) involved in mesonephric development are also involved in metanephric development (Torres et al. 1995; Kreidberg et al. 1993; Carroll et al. 2005; Kobayashi et al. 2007).

2.3
Metanephros

The development of the permanent mammalian kidney (metanephros) is initiated with the outgrowth of the UB from the caudal end of the WD. This occurs at approximately day 30 in human gestation and 10 dpc in the mouse (Woolf et al. 2003). For a period in many species, the mesonephros and metanephros co-exist (Fig. 3). The timing and site at which the UB emerges from the WD is well orchestrated such that it invades a mass of metanephric mesenchyme (MM) which will become the metanephros. Reciprocal inductive signals occur between the UB and MM (Saxen and Sariola 1987). The MM induces the UB to grow and repetitively bifurcate to form the ureteric tree via branching morphogenesis. The ureteric tree subsequently forms the collecting ducts, calyces and renal pelvis. The region of the UB that does not enter the MM becomes the ureter. Simultaneously, the epithelial cells of the tips of the ureteric tree induce committed MM cells to condense, form-

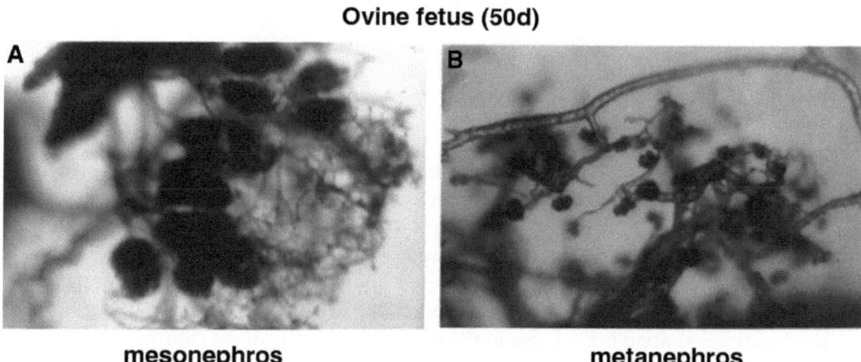

Fig. 3 Vascular casts of the sheep (A) mesonephros and (B) metanephros at 50 days of gestation. At this age, both organs co-exist. The mesonephros has relatively few, large glomeruli while, in contrast, the metanephros has an abundance of small glomeruli

ing cap mesenchyme and pretubular aggregates (Sariola 2002) (Fig. 4). The pretubular aggregates undergo MET to form renal vesicles as occurs in the development of the mesonephros. The epithelial cells of the renal vesicle develop into nephrons. This occurs through a number of stages as the renal vesicle develops firstly into a comma-shaped body followed by an S-shaped body. The upper portion of the S-shaped body develops into the distal convoluted tubule, the centre portion develops into the proximal tubule, loop of Henle and distal straight tubule, and the lower limb forms the renal corpuscle. The epithelium of the inner lining of the lower limb differentiates into glomerular podocytes, while the cells of the external portion of the lower limb become parietal epithelial cells of Bowman's capsule. The distal end of the nephron fuses with the collecting duct. Capillaries and mesangium develop in the lower cleft of the S-shaped body and are enveloped by the podocytes (Kloth et al. 1994). This layer is then incorporated around the capillary bundle. Once this

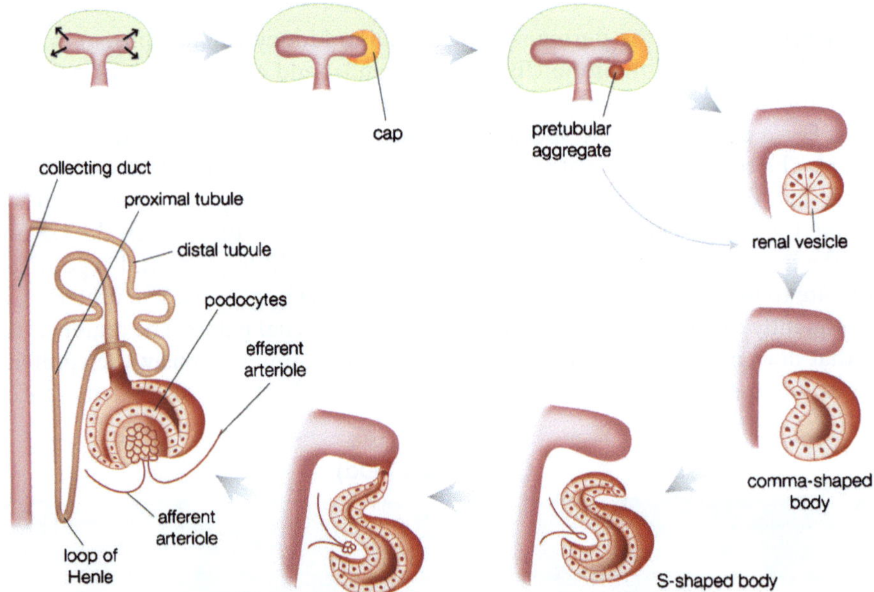

Fig. 4 Schematic representation of nephrogenesis. Signals from the ureteric epithelial tip cells induce the adjacent metanephric mesenchyme to condense forming a cap-like structure. A subset of these cells further aggregate to form the pretubular aggregates that undergo a mesenchyme-to-epithelial transformation, forming the renal vesicle. The renal vesicle further differentiates into the comma- and S-shaped bodies. Endothelial cells migrate into the cleft of the S-shaped body, which will contribute to the formation of the renal corpuscle. The upper limb of the S-shaped body fuses to the tip of the ureteric duct (future collecting duct). The lower limb of the S-shaped body gives rise to the podocytes and Bowman's capsule. The upper limb differentiates and elongates, forming the distal tubule, and the middle section forms the loop of Henle and proximal tubule. (Illustration by Ms. Nina Bosanac, Technology Services Group, Multimedia Services, Monash University)

occurs, the podocytes can no longer proliferate and they begin to differentiate producing filtration slit diaphragms and foot processes. The basement membranes of the podocytes and glomerular capillaries fuse during maturation to form the glomerular basement membrane (GBM).

In humans, early ureteric branching is involved in the generation of the renal pelvis and calyces and nephron formation does not occur (Ekblom 1992). In later generations of branches each tip induces two simultaneous nephrons. Lying adjacent to the inducing tips is a region known as the intercalated zone. This zone of growth transports the newly induced nephron to the future cortex. Nephrons are carried along with the tip from one generation to the next as new ureteric branches are formed. As the tip divides to generate two new branches the already induced nephron is carried with one tip while the other induces a new nephron. From week 15–20, the tips no longer branch but nephrogenesis continues. These nephrons do not attach to the collecting tubules but are arranged in arcades and this results in more than one nephron being attached to the same tip. This occurs in the following manner. The tip induces a nephron which connects to the tip (collecting duct). A second nephron is induced by the same tip and gradually the connecting tubule of the older nephron shifts its attachment away from the tip to the connecting tubule of the younger nephron. The tip then induces another nephron and continues this process until four to seven nephrons are found on one arcade. This process of arcade formation is shown in Fig. 5. From 20 weeks, the ampulla continues to induce nephrons, with the majority of nephrons produced in this third trimester (Ekblom 1992). These nephrons attach directly to the entire length of the collecting duct and do not become incorporated in the arcade. The pattern of bifid branching, whereby one tip induces the formation of a nephron whilst the other goes on to branch again, is thought to occur 15 times in the development of the human metanephros (al-Awqati and Goldberg 1998). If the process was 100% efficient, this would result in 32,768 nephrons. However, as discussed in Sect. 7.1, the human kidney can contain more than 2 million nephrons and this is due to the formation of nephron arcades as described above in this section. By week 34–36, nephrogenesis in humans is complete. The timing of the beginning and completion of metanephrogenesis as well as the total number of nephrons formed varies considerably between species. This is discussed below in Sect. 7.4.

2.4
Renal Vascular Development

A day after the UB has entered the MM, capillaries are detected around the perimeter of the metanephros and around the UB (Loughna et al. 1996, 1997). At 8–10 weeks gestation in the human and around 13 dpc in the mouse, capillaries can be seen forming around the S-shaped bodies. These capillaries express CD31 and vascular endothelial growth factor (VEGF) receptor. Around this same time, a single renal artery running from the dorsal aorta to the metanephros can be seen (Yuan et al. 2000). This vessel branches into smaller arteries that terminate in afferent glomerular arterioles.

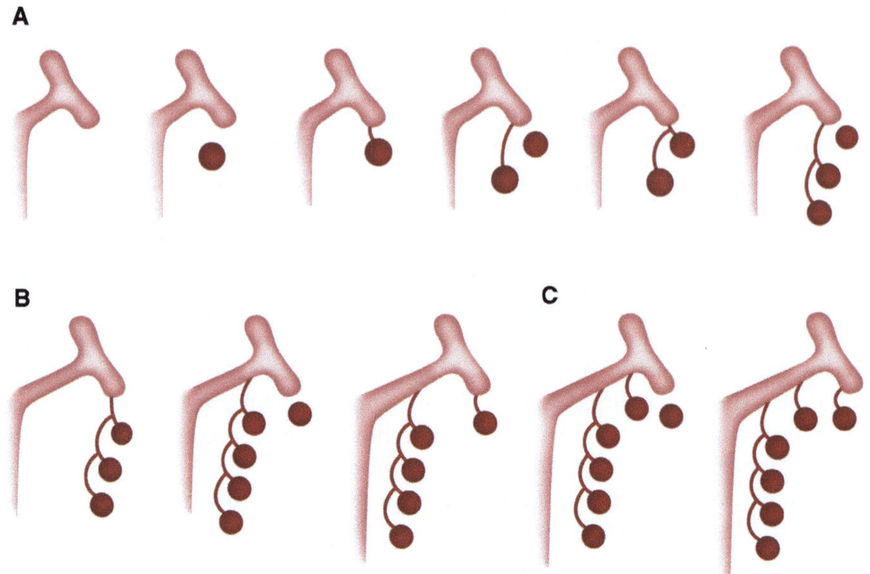

Fig. 5A–C Formation of nephron arcades in the human. From week 15–20 the tips no longer branch but nephrogenesis continues. **A** Each tip induces the surrounding metanephric mesenchyme to produce one nephron. The nephron fuses to the tip (collecting duct). **B** The same tip induces the formation of another nephron, which attaches to the tip and the connecting tubule of the older nephron shifts to the new nephron. This process continues until four to seven nephrons form an arcade. **C** After week 20, nephrons begin to attach along the entire length of the collecting duct. Figure adapted from Osathanondh and Potter 1963. (Illustration by Ms. Nina Bosanac, Technology Services Group, Multimedia Services, Monash University)

The origin of the endothelial cells that comprise the renal vasculature has been controversial. The issue has been whether the renal vasculature arises from precursor cells within the metanephros via vasculogenesis or whether they arise from existing vessels outside of the metanephros via angiogenesis. Experiments have been conducted to support both theories. The classical experiments that support the development of the renal vasculature via angiogenesis involve the transplantation of the avascular metanephros onto the quail chorioallantoic membrane. The endothelial cells that develop in the metanephros are derived from the quail host, suggesting that growth factors have attracted endothelial cells into the metanephros (Sariola et al. 1983). Experiments to support that the renal vasculature originates via vasculogenesis involved the use of a *Tie-1/LacZ* transgenic mouse to follow kidney endothelial cell development. *Tie-1* receptor tyrosine kinase is expressed on endothelial precursor cells, and cells expressing this marker were found within the avascular metanephros. In addition, other endothelial markers such as VEGFR-2 (Flk-1), Angiopoietin 1 and 2 and Tie-2 are expressed in the mesenchyme of the avascular metanephros (Loughna et al. 1997, 1998; Woolf and Loughna 1998; Yuan et al. 1999; Kolatsi-Joannou et al. 2001). However, it is still not clear whether these cells represent cells within the metanephric mesenchyme that are in the process of differentiating into endothelial cells or whether they are angioblasts that have entered the metanephros during its development. Various isoforms of PECAM-1 are involved at different stages of capillary morphogenesis in the developing mouse metanephros (Kondo et al. 2007). Renal vasculogenesis/angiogenesis can be affected by maternal undernutrition in the rat (Khorram et al. 2007).

2.5
Development of the Renal Nerves

The human kidney contains an abundance of adrenergic nerves (as determined by tyrosine hydroxylase staining) from as early as 20 weeks of gestation in both the cortex (in close proximity to renal arteries and arterioles) and medulla (close to tubular cells). Whilst the density of these receptors increases with gestation in the cortex to reach adult levels by about 28 weeks of gestation, in the medulla, receptor levels decline with increasing gestational age and were not found in the medulla of the adult (Tiniakos et al. 2004). Other nerves (staining positive for neuron-specific enolase and neurofilaments) were found in lesser abundance in the second- and third-trimester human fetal kidney (Tiniakos et al. 2004).

In the rat, renal nerves, both afferent and efferent, are present inside the kidney by 16 dpc (Liu and Barajas 1993). The afferent nerves are well developed by birth and are found within the renal pelvis, in corticomedullary connective tissue and associated with the renal vasculature (Liu and Barajas 1993). Efferent nerves are less well developed at birth and are found in proximity of interlobular arteries and the afferent arterioles of the juxtamedullary nephrons. However, after birth, the efferent nerves grow rapidly and achieve a distribution similar to the adult by day 21 after birth (Liu and Barajas 1993). Overall, it appears the afferent renal innervation precedes that of the efferent innervation in the developing kidney, suggesting an important role for growth and development.

The renal nerves have been shown in the sheep to play a role in regulating renal function during development (reviewed in Robillard et al. 1993). During fetal life, the renal nerves play a role in regulating renin secretion (Ito et al. 2001), especially during the transition from fetus to newborn (Page et al. 1992), whilst adrenergic receptor blockade in the fetus causes renal vasodilation (Robillard et al. 1993). However, fetal renal denervation in the last third of gestation does not alter fetal glomerular filtration rate (GFR) or renal blood flow (Smith et al. 1990).

3
Genetic Regulation of Metanephric Development

Molecular regulation of renal development has been an area of intense study over the last decade. Below we have highlighted critical studies demonstrating the importance of specific genes and gene families at particular stages of renal development;

however, the list is by no means complete and readers are referred to some recent reviews which deal with this subject in greater depth (Clark and Bertram 1999; Clark et al. 2001; Pohl et al. 2000; Davies 2001; Bouchard 2004; Cullen-McEwen et al. 2005; Costantini and Shakya 2006; Schmidt-Ott et al. 2006; Boyle and de Caestecker 2006).

3.1
Molecular Specification of the Metanephric Blastema

One of the first steps in the development of the metanephros is the specification of the metanephric blastema from the intermediate mesoderm located at the caudal end of the nephrogenic cord at around 10–10.5 dpc in the mouse. Several genes are expressed in the metanephric mesenchyme before UB outgrowth, including *Odd-1* (James et al. 2006), *Eya1* (Kalatzis et al. 1998; Xu et al. 1999), *Pax2* (Dressler et al. 1990; Torres et al. 1995), *Wt-1* (Kreidberg et al. 1993), *Six1* (Xu et al. 2003), *Gdnf* (Moore et al. 1996; Pichel et al. 1996; Sanchez et al. 1996) and *Sall1* (Nishinakamura et al. 2001). However, apart from *Eya1* and *Odd-1* the loss of function of all these genes still results in the expression of a subset of metanephric mesenchyme. Thus, these results indicate that to date *Eya-1* and *Odd-1* are the earliest markers required for the specification of the metanephric blastema (Sajithlal et al. 2005; James et al. 2006).

3.2
Molecular Regulation of Ureteric Budding and Branching Morphogenesis

In addition to the specification of the metanephric blastema, the development of the metanephros requires the formation of the nephric duct/WD. Loss of function of genes expressed in the WD such as *Lim1* (Shawlot and Behringer 1995; Tsang et al. 2000), *Pax2* (Dressler et al. 1990; Torres et al. 1995) and *Pax8* (Bouchard et al. 2002) result in the disruption of nephric duct formation and subsequently renal agenesis.

The next step in the development of the metanephros is the outgrowth of the UB from the WD. This involves the interplay between two major transforming growth factor-β (TGF-β) superfamily signalling pathways: the bone morphogenetic protein 4 (BMP4) and glial cell line-derived neurotrophic factor (GDNF) pathways. GDNF is expressed in the intermediate mesoderm (presumptive metanephric blastema) adjacent to the caudal region of the WD (Hellmich et al. 1996). GDNF signals through the c-Ret receptor tyrosine kinase and the GPI-linked co-receptor, Gfra1. Both of these receptors are expressed by the WD. Upon branching of the ureteric tree, c-Ret becomes restricted to the tips of the ureteric tree and GDNF to the mesenchyme surrounding these tips, allowing continued induction of branching to occur only at the tips. Mice lacking GDNF, c-Ret or Gfra1 display renal agenesis due to lack of budding or blind-ended ureters or small disorganized kidney rudiments (Durbec et al. 1996; Moore et al. 1996; Pichel et al. 1996; Sanchez et al. 1996; Schuchardt et al. 1996; Sainio et al. 1997; Cacalano et al. 1998; Enomoto et al. 1998; Sariola and Saarma 1999). Figure 6 shows the role of major genes involved in early budding and branching of the metanephros.

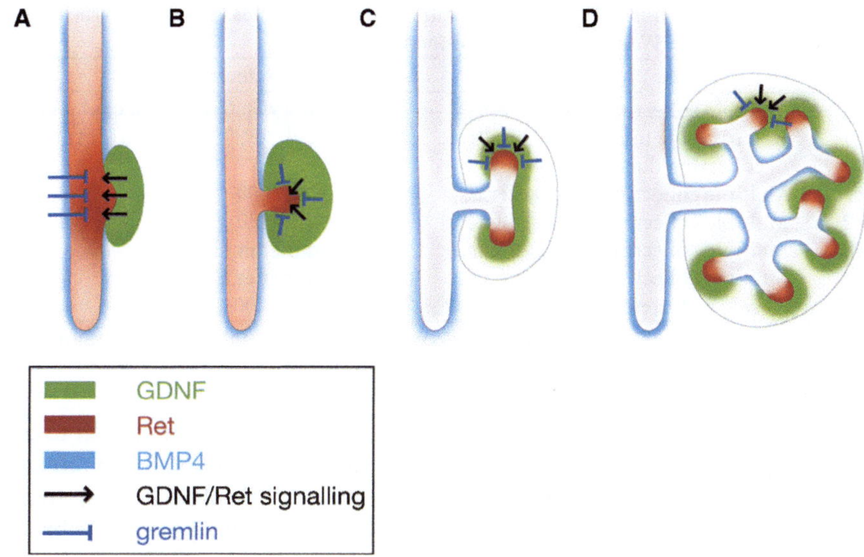

Fig. 6A–D The expression of regulators of ureteric budding and branching morphogenesis. **A, B** c-Ret is expressed throughout the Wolffian duct prior to ureteric budding but becomes more highly expressed at the bud site. BMP4 is expressed in the mesenchyme adjacent to the Wolffian duct, inhibiting ectopic budding from occurring. GDNF is expressed throughout the metanephric mesenchyme. Gremlin antagonizes the action of BMP4 at the bud site whilst GDNF interacts with c-Ret at the bud site to promote bud outgrowth. **C, D** As c-Ret becomes restricted to the ureteric epithelium tips, GDNF also becomes restricted to the mesenchyme surrounding these tips in the nephrogenic zone. Gremlin antagonizes the action of BMP4 at the tips, allowing branching to occur. c-Ret/GDNF signalling and BMP4/gremlin signalling regulate subsequent branching events. (Adapted from Costantini and Shakya 2006 and Michos et al. 2007. Illustration by Ms Nina Bosanac, Technology Services Group, Multimedia Services, Monash University)

The precise emergence site of the UB from the WD involves the interplay of several genes that have been shown to promote the expression of GDNF (*Pax2*, *Eya1*, *Sall1* and all *Hox11* paralogs), regulate the GDNF expression domain (*Foxc1/2*, *BMP4* and *Slit2/Robo2*) or the response to GDNF (*Spry 1*) (reviewed in Bouchard 2004; Costantini and Shakya 2006). Genes that regulate the expression domain of GDNF such as *FoxC1/2*, *BMP4* and *Slit2/Robo2* and the activity of BMP4, such as *gremlin*, play important roles in determining the position of the bud site from the WD. This is not surprising given that the majority of these genes are strongly expressed in the mesenchyme that surrounds the WD (nephrogenic cord) and/or the UB (Fig. 6). As shown in Table 1, mice with a disruption in any of these genes display phenotypes that arise from ectopic budding of the UB from the WD or ectopic budding from the UB itself (Kume et al. 2000; Miyazaki et al. 2000; Grieshammer et al. 2004).

Table 1 Urinary tract abnormalities in mice with mutations in genes expressed in the mesenchyme surrounding the Wolffian duct (nephrogenic cord)

Gene	Mouse phenotype	Reference
AT2	Multiple collecting duct systems	Pope et al. 1998; Nishimura et al. 1999
Bmp4	Heterozygote mice exhibit multiple collecting duct systems	Dunn et al. 1997; Miyazaki et al. 2000
Foxc1	Multiple collecting duct systems	Kume et al. 2000
Foxc2	Multiple collecting duct systems	Kume et al. 2000
Robo2	Multiple collecting duct systems	Grieshammer et al. 2004
Slit2[a]	Multiple collecting duct systems	Grieshammer et al. 2004
Grem1	Absence of ureters and renal agenesis	Michos et al. 2004, 2007

[a] *Slit2* is expressed weakly in the nephrogenic cord and more strongly in the Wolffian duct

Once the UB enters the MM, several other molecules have been shown to promote branching (e.g. hepatocyte growth factor (HGF), heparan sulphate proteoglycans, integrins α3 and α8, matrix metalloproteinase-9, Erk MAP-kinase), inhibit branching (e.g. activin, TGF-β2, BMP-2, protein kinase A pathway), regulate branch elongation (TGFβ1 and HGF) (reviewed in Clark and Bertram 1999; Clark et al. 2001; Martinez et al. 2001; Pohl et al. 2000; Davies 2001), or control branch symmetry (*Hoxa11*, *Hoxd11* and *BMP4*) (Patterson et al. 2001; Raatikainen-Ahokas et al. 2000; Cain et al. 2005). Many of these molecules demonstrate a temporal–spatial expression pattern in the MM and/or UB.

3.3
Molecular Regulation of Nephrogenesis

The first step of nephrogenesis involves the epithelialization of the uninduced MM through factors secreted by the UB. These include leukaemia inhibitory factor (LIF), TGF-β2, lipocalin-2, Wnt-6, BMP7 and cytokine receptor-like factor-1 (CRLF-1) (Perantoni et al. 1995; Barasch et al. 1999; Plisov et al. 2001; Itäranta et al. 2002; Sariola 2002; Schmidt-Ott et al. 2005, 2006). The cap mesenchyme expresses a number of transcription factors, such as *Pax-2*, *Cited-1*, *Six1*, *Sall1* and *Eya1* (reviewed in Schmidt-Ott et al. 2006; Boyle and de Caestecker 2006). The cells of the pretubular aggregate which develop into the nephron express Wnt-4 and Lim-1 and begin to express epithelial markers (Sariola 2002; Barasch et al. 1999; Plisov et al. 2001; Schmidt-Ott et al. 2006) as MET begins to take place and the renal vesicle develops. As nephrogenesis proceeds, genes become spatially restricted to various segments of the nephron. These include members of the cadherin family (Dahl et al. 2002; Cho et al. 1998), Notch signalling pathway (Chen and al-Awqati 2005; Piscione et al. 2004), Robo/Slit family (Piper et al. 2000) and aquaporins (Liu and Wintour 2005), just to name a few.

A number of genes have been identified to be expressed in the developing podocytes of the glomerulus. These include *Wt-1*, *podocalyxin*, *nephrin (Nphs1)*, *podocin (Nphs2)*, *Pod1*, *synaptopodin*, *LMX1B* (Chen et al. 1998) and *protein tyrosine phosphatase receptor*. Mutations in many of these are associated with glomerular disorders in mice and humans demonstrating the importance of podocytes in the function of the glomerular filtration barrier (reviewed in Kreidberg 2003; Pätäri-Sampo et al. 2006; Rascle et al. 2007).

Table 2 Genes involved in Kidney development and anomalies caused by gene deletion in mice[a]

Gene	Renal mouse phenotype	Reference
Ace	Cortical thinning, focal areas of atrophy, vascular thickening; hypertensive	Krege et al. 1995
Agt	Hypoplastic renal papilla and widening of the renal pelvis region; renal cysts; delay in glomerular maturation; hypertensive	Niimura et al. 1995
AT1a & b	Vascular thickening within the kidney and atrophy of the inner renal medulla; hypertensive	Oliverio et al. 1998
Bmp7	Renal dysplasia	Dudley et al. 1995
R-Cad	Culture of R-Cdh$^{-/-}$ embryonic kidneys revealed a 35% reduction in the ratio of nephron number to ureteric tip number; no change in number of UB tips.	Dahl et al. 2002
Cad–6	Delay in the formation of renal vesicles and failure of some renal vesicles to connect to the collecting duct	Mah et al. 2000
Emx2	Bilateral agenesis due to failure of UB branching	Miyamoto et al. 1997
fgfr2IIIb	Hypoplasia	Revest et al. 2001
Fgf10	Hypoplasia	Ohuchi et al. 2000
Frem1	Renal agenesis (20% unilateral agenesis in mice)	Smyth et al. 2004; Kiyozumi et al. 2006
Gdnf	Renal agenesis (73% bilateral, 27% have unilateral hypoplastic kidney remnant) due to UB defects; heterozygotes have 30% fewer nephrons	Moore et al. 1996; Pichel et al. 1996; Sainio et al. 1997; Cullen-McEwen et al. 2001
Gfrα1	Renal agenesis (76% bilateral, 24% contain a unilateral hypoplastic kidney remnant)	Cacalano et al. 1998
Grip1	Renal agenesis	Takamiya et al. 2004
Gdf 11	Renal agenesis	Esquela and Lee 2003
Hs2st	Renal agenesis (100% bilateral)	Bullock et al. 1998
Hoxa11/ Hoxd11	Rudimentary or absent kidneys; defects in UB branching; control development of a dorsoventral renal axis	Patterson et al. 2001
Lama5	Renal agenesis (in a few mice); 20% of mice demonstrate no branching after initial budding	Miner and Li 2000
Lim1	Bilateral agenesis	Shawlot and Beringer 1995
Mmp14	Kidneys contain poorly differentiated tubules	Kanwar et al. 1999
Notch2[b]	Hypoplastic kidneys; mutant glomeruli lacked a normal capillary tuft	McCright et al. 2001
Odd–1	Bilateral agenesis due to lack of metanephric mesenchyme	James et al. 2006

(continued)

Table 2 (continued)

Gene	Renal mouse phenotype	Reference
Osr1	Renal agenesis	Wang et al. 2005
Pax8	Renal agenesis	Bouchard et al. 2002
Ret	Renal agenesis (58% bilateral, 31% unilateral)	Liu et al. 1996; Schuchardt et al. 1996
Shh	Renal hypoplasia; hydroureter, dilated pelvis	Yu et al. 2002
Six 1	Bilateral agenesis	Xu et al. 2003
Six2	Renal hypoplasia	Self et al. 2006
Slit3	Unilateral or bilateral agenesis of the kidney and ureter or varying degrees of renal hypoplasia	Liu et al. 2003
Spry1	Multiple collecting duct systems	Basson et al. 2005
Tgfb2	Dilated renal pelvis, hypoplastic and cysts	Sanford et al. 1997
Wnt-4	Small, dysgenic kidneys; lack pretubular aggregates	Stark et al. 1994
Wnt-11	30% nephron deficit	Majumdar et al. 2003

[a] Table does not include those genes mentioned in other tables
[b] Notch2 mutant is a hypomorph
UB, ureteric bud

3.4
Molecular Regulation of the Stroma

Stromal cells are first evident peripheral to the cap mesenchyme cells and express the transcription factor, *Foxd1* (formerly *BF-2*) (Hatini et al. 1996). Once nephrogenesis has been initiated and several rounds of branching have taken place stromal cells are found arranged around the ureteric branches and developing nephrons and are often referred to as primary renal interstitium (Alcorn et al. 1999). These cells express glycolipid disialoganglioside (GD3) (Sariola et al. 1988), tenascin (Ekblom and Weller 1991) and Foxd1 (Hatini et al. 1996). By late gestation two distinct stromal populations arise from the cortical and medullary stroma. The renal stroma provides a structural framework around the developing nephrons and collecting duct system. In the mouse, inactivation of a number of genes expressed in the renal stroma has demonstrated the importance of this renal sub-compartment in regulating both branching morphogenesis and nephrogenesis (reviewed by Cullen-McEwen et al. 2005; see Table 3).

3.5
The Renin–Angiotensin System: An Important System Regulating Renal Development and Function

It is worth noting here the importance of an intact renin–angiotensin system (RAS) for normal renal development (recently reviewed in Lasaitiene et al. 2006) because as discussed later, altered gene expression of components of this system are found in models where kidney development is impaired due to a perturbation in the maternal environment. The most active peptide of this system, angiotensin II (Ang

Table 3 Renal abnormalities in mice with mutations of genes expressed in the renal stroma

Gene	Mouse phenotype	Reference
Fgf7	Kidneys have 30% fewer nephrons, hypoplastic papilla and cysts	Qiao et al. 1999b
Foxd1	Abnormal collecting duct system, renal hypoplasia and severe defects in nephrogenesis	Hatini et al. 1996
Pbx1	Renal hypoplasia, reduced nephron number, defects in ureteric branching; the kidneys are rotated ventrally and positioned caudally to normal	Schnabel et al. 2003
Rarα/Rarβ	RARα and RARβ double knock-outs have a fourfold decrease in UB tips and renal hypoplasia	Mendelsohn et al. 1999
Pod1	Hypoplasia with a 61% decrease in branching	Quaggin et al. 1999

UB, uteric bud

II), can act on both the angiotensin type 1 (AT1) and type 2 (AT2) receptors, both of which are present in abundance in the developing metanephros. In the rodent, there are two subtypes of the AT1 receptor (AT1a and AT1b). Ang II binding to these receptors can have a wide range of effects. Ang II via the AT2 receptor causes upregulation of Pax-2, which may contribute to MET, tubular proliferation as well as mediating apoptosis (Zhang et al. 2004a, 2004b). In contrast, in renomedullary interstitial cells, Ang II via the AT2 receptor has antiproliferative actions (Maric et al. 1998). Absence of the AT2 receptor results in congenital abnormalities of the kidney and urinary tract (discussed below in Sect. 4.2) due to delayed apoptosis of mesenchymal cells (Nishimura et al. 1999).

Ang II acting on the AT1 receptor plays a crucial role in tubular development where it mediates growth and proliferation of proximal tubules and loops of Henle (Wolf and Nielsen 1993). It has also been shown that the AT1 receptor has a role in branching morphogenesis (Iosipiv and Schroeder 2003). During nephrogenesis, lack of AT1 receptor activation can inhibit E-cadherin (Lasaitiene et al. 2003) and integrin α6 (Chen et al. 2004) and thus disrupt establishment of epithelial polarity and cell–cell interactions. Ang II also has important roles in nephrovascular development (Tufro-McReddi et al. 1995). Given the wide range of effects mediated by the AT1 receptor, it is not surprising that absence of the AT1 receptor (double AT1a and AT1b knock-out) or pharmacological inhibition of the RAS results in severe renal abnormalities, including renal tubular malformations, medullary atrophy and impairment of urine concentrating ability (Niimura et al. 1995; reviewed in Lasaitiene et al. 2006).

3.6
Summary of Molecular Regulation of Metanephric Development

In this section, we have described the importance of a number of key genes that play a role in the development of the metanephros. Much of what we know thus far has been obtained through the analysis of kidney phenotypes displayed by knock-out mice (Tables 1–3). In recent years, there has been an explosion in the number of genes known to be temporally and spatially expressed in the developing metanephros through the analysis of global gene expression (see Sect. 5.4). Many of these genes await functional analysis to determine their roles in the development of the metanephros.

4
Abnormalities of Renal Development in the Human

Many animal models have been developed to study the aetiology of fetal renal disease (Peters 2001; Chevalier 2004). Apart from the immediate effects on the functioning of the fetal and neonatal kidney, one also has to consider the potential permanent alterations brought about in the metanephros, and the likelihood that these changes will lead to accelerated renal damage with aging and increase the risk of cardiovascular disease (Woolf 2001; Eskild-Jensen et al. 2002; Chevalier 2004).

4.1
Hydronephrosis

Hydronephrosis is one of the most common problems detected (by prenatal ultrasound) in the developing kidney and has a prevalence at birth of 0.5%–4.5% (Walsh et al. 2007). Infants with hydronephrosis are nearly 12 times more likely to be hospitalized in the first year of life with pyelonephritis-related problems.

4.2
Congenital Anomalies of the Kidney and Urinary Tract

Congenital anomalies of the kidney and urinary tract (CAKUT) is a clinical description of complex developmental renal and ureteric abnormalities that have been recognized within families. CAKUT accounts for one-third of all anomalies detected by routine fetal ultrasound (Woolf et al. 2004; Noia et al. 1989). They include renal agenesis, hypoplasia/dysplasia, multicystic dysplastic or duplex kidney often associated with vesicoureteric reflux (VUR), hydroureter, hydronephrosis or obstruction at the vesicoureteric (VUJ) or ureteropelvic (UPJ) junction. These anomalies occur in various combinations within families, suggesting an incomplete or variable genetic penetrance. Severity may vary from incidental clinical findings to chronic ill health and end-stage renal failure in childhood. CAKUT often occurs as part of a syndrome with multiple developmental anomalies, and although the genetic mutations responsible for some of these syndromes have been identified (Tables 1 and 2), there are many more in which the causative gene is still unknown (Pope et al. 1999; Miyazaki et al. 2003; Woolf et al. 2004; Woolf 2006).

The identification of the key developmental events that are perturbed in patients presenting with CAKUT has come about through the study of several mouse models listed in Tables 1 and 2. The developmental events that are perturbed, resulting in the diverse range of anomalies, include ectopic ureteric budding from the WD leading to inappropriate insertion into the MM and/or inappropriate insertion into the bladder, the lack of ureteric budding resulting in renal agenesis, defects in branching morphogenesis, defects in the survival and/or differentiation of the MM and smooth muscle cell abnormalities in the ureter leading to peristalsis defects (Pope et al. 1999; Miyazaki et al. 2003; Woolf et al. 2004; Miyazaki et al. 1998; Airik et al. 2006; Mahoney et al. 2006).

4.3
Bilateral Renal Agenesis (Potter's Syndrome)

Failure of the ureteric bud to induce development in the metanephric blastema results in renal agenesis. Although babies can survive in utero with no kidneys, no urine is produced and infants die within hours of birth from a respiratory failure caused by pulmonary hypoplasia (Potter 1965). The lack of amniotic fluid (oligohydramnios) causes compression of the fetus, resulting in fetal deformations such as those seen in Potter's syndrome.

4.4
Renal Abnormalities Caused by Maternal Drug Use

Renal abnormalities in the neonate have been found following maternal exposure to pharmacological agents used for medicinal purposes. The two most common ones are maternal exposure to angiotensin converting enzyme (ACE) inhibitors (used for controlling hypertension) and prostaglandin inhibitors such as indomethacin (used to treat polyhydramnios). Treatment with ACE inhibitors and COX-1 inhibitors (indomethacin) during pregnancy often leads to oligohydramnios (lack of amniotic fluid) due to fetal anuria (Sawdy et al. 2003) and results in lung abnormalities after birth (Shotan et al. 1994; Kirshon et al. 1988). These drugs should be avoided in pregnancy.

4.5
Polycystic Kidney Disease

Autosomal dominant polycystic kidney disease (ADPKD) is a life-threatening genetic disease in which cysts develop primarily in the kidneys. Mutations in two particular genes, *PKD1* and *PKD2*, are the underlying cause of ADPKD, with the majority of cases (80%–85%) involving *PKD1* (Al-Bhalal and Akhtar 2005). The *PKD1* gene encodes for polycystin 1, a protein involved with cell–cell and cell–matrix interactions. Both *PKD1* and *PKD2* are expressed in the developing kidney. In mice with loss-of-function mutations of *PKD1* or *PKD2*, renal development is normal until day 15, but at this time renal cysts begin to form (Watnick and Germino 1999).

5
Methodology to Examine Kidney Development

5.1
Organ Culture

Whole metanephric organ culture, developed by Clifford Grobstein in the 1950s (Grobstein 1956) has been the most important in vitro model system to date for studying kidney development. With this technique, whole rat or mouse metanephroi are cultured in defined media or in the presence of serum from approximately 13.5 or 11.5 dpc, respectively, when the ureteric tree contains just a few branches and the kidney contains no mature nephrons. The metanephroi are cultured at an air–media interface on Transwell filters for up to approximately 6 days, by which time they typically contain at least 50 glomeruli at different stages of maturation, and at least this number of ureteric branches. Developing glomeruli can be visualized and quantified by whole-mount immunostaining the cultures with WT-1, which is a marker of developing podocytes. The development of the ureteric tree can be visualized by immunostaining with calbindin-D_{28K}-or cytokeratin (Fig. 7). Alternatively, metanephroi from *Hoxb7/GFP* transgenic mice in which the entire ureteric epithelium expresses green fluorescent protein (GFP) can be cultured to monitor the development of the ureteric epithelium (see Sect. 5.5 below).

Using this methodology, researchers can manipulate metanephric organ cultures to determine the effects of exogenously added growth factors and hormones in the media or localized on agarose beads (see Fig. 8) (Vilar et al. 1996; Sainio et al. 1997; Qiao et al. 1999b; Raatikainen-Ahokas et al. 2000; Cain et al.

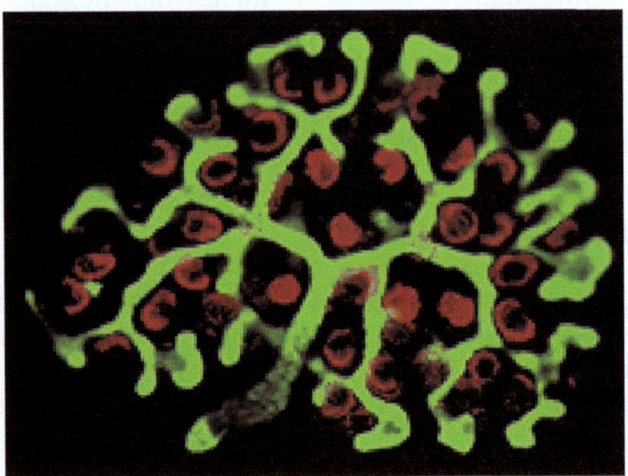

Fig. 7 Photomicrograph of a cultured mouse metanephros that has been double immunolabelled with anti-calbindin-D_{28K} (*green*) marking the ureteric epithelium and anti-WT-1 marking the developing glomeruli (*red*). (Figure courtesy of Mr. Kenneth Walker)

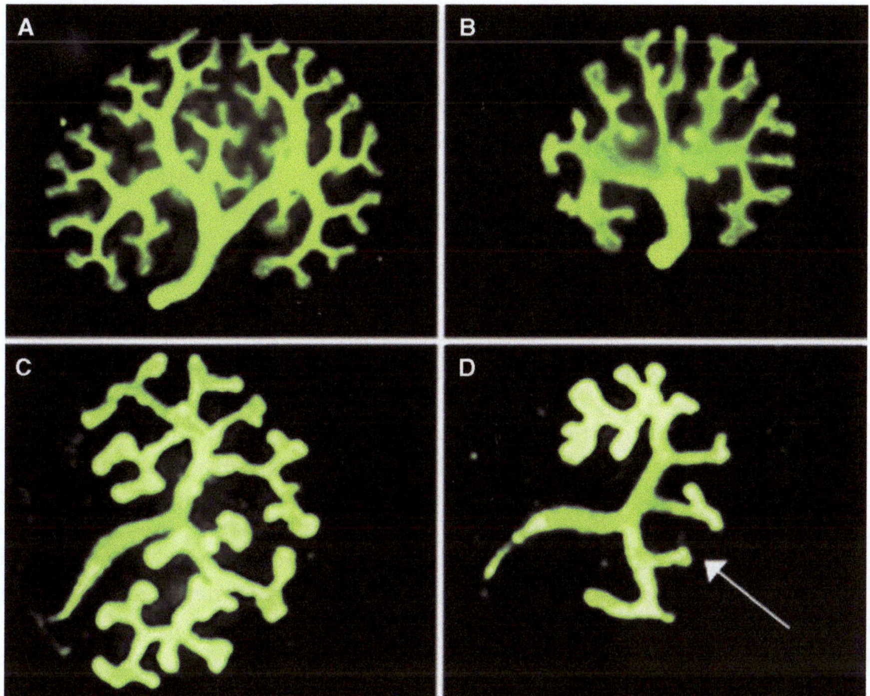

Fig. 8A–D Photomicrographs of cultured metanephroi cultured in dexamethasone or BMP4. **A** Rat metanephroi cultured at 13.5 dpc for 48 h in control media or **B** media supplemented with dexamethasone (10^{-5} M). **C** Mouse metanephroi cultured at 12.5 dpc for 48 h in control media or **D** media supplemented with 260 ng/ml recombinant BMP4. Both dexamethasone and BMP4 inhibit ureteric branching with BMP4, demonstrating a predominant inhibition of branching in the posterior region of the kidney (*arrow*). Mouse metanephroi are immunolabelled with calbindin-D_{28K} whilst the rat metanephroi cytokeratin stained. (Figures courtesy of Ms. Reetu Singh and Dr. Jason Cain)

2005; Singh et al. 2007b), inhibiting antibodies (Sorokin et al. 1990; Pugh et al. 1995) and anti-sense morpholinos or small interference RNAs (Sariola et al. 1991; Sainio et al. 1994; Quaggin et al. 1997; Li et al. 2000; Davies et al. 2004; Maeshima et al. 2006). In addition, metanephroi from genetically manipulated mice can be cultured and metanephric development analysed (Srinivas et al. 1999b; Naim et al. 2005; Sajithlal et al. 2005; Levinson et al. 2005; Cain and Bertram 2006; Self et al. 2006). It is also possible to study ureteric budding from the Wolffian duct in vitro by culturing mouse urogenital tracts at 10–10.5 dpc, prior to the emergence of the UB (Maeshima et al. 2006).

In addition to the culture of whole urogenital tracts or metanephroi, it is possible to culture dissected MM and UBs in a recombined fashion. This experiment typically involves recombining wild-type tissue with tissue obtained from a genetically altered mouse. Given that reciprocal inductive signals are sent between the ureteric epithelium and MM, it is not always evident which tissue compartment is contributing to a particular phenotype. Thus, recombination experiments allow the investigator to determine which tissue is defective. MM and UBs can also be cultured in isolation but must be supplemented with appropriate growth factors in order for nephrogenesis or branching to occur, respectively. Isolated MM is cultured in a similar way to whole metanephroi; however, isolated UBs must be cultured in a collagen gel matrix or Matrigel in order to maintain their 3D tubular structure (Barasch et al. 1999; Karavanova et al. 1996; Qiao et al. 1999a).

5.2
Three-Dimensional and Four-Dimensional Imaging

We have developed a method for measuring the length of individual ureteric branches and thereby the total length of the ureteric tree in cultured mouse metanephroi in 3D (Cullen-McEwen et al. 2002; Harper et al. 2001; Fricout et al. 2001, 2002; Cain et al. 2005). This technique was originally established in the laboratory of Dr Frank Costantini (Srinivas et al. 1999a; Watanabe and Costantini 2004; Shakya et al. 2005) and involves culturing metanephroi in an environmental chamber fitted to a confocal microscope which is controlled for temperature, humidity, pH and sterility. The metanephroi are then whole-mount calbindin-D_{28k} immunolabelled (see Sect. 5.1) or kidneys from *Hoxb7/GFP* mice (see Sect. 5.5) are used. Computer-based image segmentation, skeletonization and measurements is then performed. The algorithm performs semi-automatic segmentation of a set of confocal images and automatic skeletonization of the resulting binary object. Length measurements and number of branch points are automatically obtained. The final representation can be reconstructed, providing a fully rotating 3D perspective of the skeletonized tree. For full details see Cullen-McEwen et al. (2002). Using this technique, we found that after 36 h culture of 12.5 dpc mouse metanephroi, the total length of the ureteric tree was 6103 ± 291 μm (mean ± SD), a fourfold increase compared with metanephroi cultured for just 6 h (1522 ± 149 μm) (Cullen-McEwen et al. 2002). Ureteric duct length increased at a rate of 153 μm/h over the first 30-h period and was maximal between 18 and 24 h at 325 μm/h. The distribution of branch lengths at the six time points studied was similar, suggesting tight control of the pattern of ureteric lengthening and branching.

Using the *Hoxb7/GFP* reporter mice, we have been able to image the same kidney on multiple occasions, and thereby image the development of the ureteric tree over time (4D). We have cultured metanephroi from 12.5 dpc *Hoxb7/GFP* mice on a Leica TCS-NT confocal microscope for up to 72 h with confocal images captured every 6 h. A time-lapse movie of an 12.5 dpc *Hoxb7/GFP* transgenic mouse kidney cultured at the confocal microscope for 72 h and imaged every 2 h can be seen at http://www.med.monash.edu.au/anatomy/research/ kidneydevelopment.html. As more reporter mice are generated (see Sect. 5.5), this will enable the visualization and quantitation of defined cell populations during development.

5.3
Quantitation of Nephron Number

The total number of nephrons (glomeruli) in a kidney has emerged in the past 10–15 years as an important index of renal structure and a potentially important index of renal health. This relatively recent interest in nephron number has emerged for a number of reasons: (1) the hypothesis of Brenner et al. (1988) linking low nephron number and hypertension (if nephron number was similar in all kidneys of a given species, then the Brenner hypothesis would be untenable, See Sect. 8.5.1 below); (2) the development of unbiased stereological methods for estimating (counting) the total number of nephrons in kidneys; and (3) the findings that a variety of environmental and genetic factors regulate nephrogenesis and thereby nephron endowment during metanephric development.

5.3.1
Unbiased Counting of Nephrons

Stereology has been defined as the discipline concerned with the quantitative analysis of three-dimensional structures (Bertram 1995). Biologists have used stereological methods for more than a century to measure objects of interest. Typically, this involves macroscopic sampling of the tissue or organ of interest, the generation of a set of histological and/or electron microscopic sections and measurement of features on the sections. Unfortunately, while unbiased stereological methods have been available for estimating the absolute and relative volumes, surface areas and lengths of tissue components of interest for decades (see Weibel 1979), the methods available to count objects were limited, as described in this section.

The stereological counting methods that were available until the early 1980s were limited for two major reasons. First, assumptions were required of the shape, size and/or size distribution of the objects being counted. For example, when counting glomeruli (and thereby nephrons), a priori knowledge of glomerular shape, size and/or size distribution was required. Typically, this knowledge was not available and therefore assumptions of glomerular geometry were required. These methods are known as model-based stereological methods, because they required knowledge (usually replaced by assumptions) of the geometry (geometric model) of glomerular size and shape. To the extent that these model assumptions deviated from the truth, then the final estimation of number was biased. These methods included those of Abercrombie (1946), Floderus (1944), Weibel and Gomez (1962) and DeHoff and Rhines (1961).

The second major problem associated with the model-based methods of stereological counting was that the final estimates were typically expressed in terms of numerical density (N_V) rather than total number. The literature is replete with reports of glomerular density, such as the number of glomeruli per unit volume of cortex (typically abbreviated $N_{Vglom,cortex}$) or the number of glomeruli per unit volume of kidney ($N_{Vglom,kidney}$). While these numerical density estimates provide information on the number of glomeruli within a unit volume, they of course do not provide any indication of the total number of glomeruli (nephrons) per kidney.

An additional limitation of numerical density estimates is the problem of the so-called reference trap. This refers to the problem of dimensional changes in the reference volume during preparation of tissue for microscopic analysis. To the extent that the tissue (say renal cortex) swells or shrinks during fixation, processing, embedding and sectioning, then the value of N_V will also change. If the degree of dimensional changes in the reference space differs between specimens or experimental groups, then the observed differences in N_V likely tell us more about tissue shrinkage than about the number of glomeruli. This clearly can result in major errors in data interpretation.

Any consideration of glomerular counting would be incomplete without a brief consideration (condemnation) of the practice of expressing glomerular number in terms of the number per unit area ($N_{Aglom,cortex}$) of cortex. Such estimates are obtained by counting the number of glomerular profiles (in stereology a profile is defined as a two-dimensional representation of a particle (e.g. glomerulus) once it has been sectioned) observed per unit area of sectioned cortex (usually histological sections). This is easy data to obtain, but the dangers associated with interpreting NV estimates (as described above in this section) apply equally to the interpretation of NA estimates. However, the problems are compounded because the likelihood of a glomerulus being contained in a histological section (i.e. sampled by a section) again depends very much on the size and shape of that particular glomerulus. Given that the volume of individual glomeruli varies widely in kidneys (Samuel et al. 2005), and glomeruli can hypertrophy or shrink under different pathological circumstances, N_A differences between specimens may well tell us more about differences between the specimens in glomerular size than in glomerular number. Unfortunately, differences in N_A between specimens are almost always mistakenly interpreted as differences in total glomerular number, which can again lead to serious misinterpretation of experimental outcomes.

The publication of the disector method (Sterio 1984) revolutionized stereology and led to the development of a new generation of unbiased stereological techniques, including techniques for estimating the total number of glomeruli (nephrons) in a kidney (i.e. counting nephrons). The disector is essentially a sampling tool that samples particles (three-dimensional objects such as glomeruli) in three-dimensional space (such as kidneys) with equal probability. In other words, all glomeruli have the same chance of being sampled, and subsequently counted, regardless of their size, shape or location in the kidney. The disector thus overcomes the problem discussed at the beginning of this section that glomeruli within kidneys (and between kidneys) have different sizes, size distributions and shapes. With the disector, glomeruli are sampled according to their number, not their geometric characteristics. Counting strategies based on the disector principle are thus said to be unbiased because all glomeruli (nephrons) have an identical chance of being sampled and subsequently counted.

Two general approaches are available for estimating the total number of glomeruli, and thereby nephrons, in the kidney using the disector principle. These are the physical disector/fractionator approach and the physical disector/Cavalieri approach. Both are described in detail in Bertram (1995, 2001) and Nyengaard and Bendtsen (1992).

With the more commonly used physical disector/fractionator method, physical disectors are used to count glomeruli in a known fraction of the kidneys. This known fraction of kidney tissue is obtained via a series of macroscopic and microscopic sampling steps. First, slicing devices are typically used to obtain macroscopic slices (and sub-slices if large kidneys are being analysed) of kidneys, and then a known fraction of these kidney slices and sub-slices is embedded in glycolmethacrylate, an embedding medium that undergoes minimal dimensional changes. The embedded tissue samples are then exhaustively serially sectioned (until no tissue remains in the block), and a known fraction of the sections is collected and mounted on glass slides. During sectioning, pairs of sections are collected, for example, adjacent sections. These pairs of sections must then be viewed simultaneously. This can be achieved using pairs of microscopes fitted with projection arms (see Bertram 2001) or using a split-screen approach as provided by several commercially available systems (CASTGrid System, Olympus; Stereo Investigator, MicroBrightField Inc.) Corresponding fields on the pairs of sections are found and those glomeruli sampled by an unbiased counting frame on the field of the first section that are not present in the corresponding field of the second section are counted. To double the efficiency of the technique, those glomeruli sampled by an unbiased counting frame in the field of the second section, that are not present in the first section, are also counted. Again, the fraction of the section area examined in this way must be known. To calculate total glomerular number in a kidney with the physical disector/fractionator approach, the actual number of glomeruli counted (typically between 100 and 200 per kidney) is multiplied by the reciprocals of the various sampling fractions (slice sample fraction, section sampling fraction, field sampling fraction), to provide an unbiased estimate of total glomerular number.

To count glomeruli with the physical disector/Cavalieri combination, physical disectors are used to determine glomerular numerical density (number per volume), and this is multiplied by the volume of the kidney or cortex (depending on which reference space is being used), which is estimated using the Cavalieri principle (see Bertram 1995, 2001). The Cavalieri principle can be used to estimate the volume of any object, regardless of its shape or size. All that is required is a set of sections through the object of similar thickness and separation. For example, to estimate the volume of an adult rat kidney, an exhaustive set of 1-mm slices is obtained (usually about 16), the total area of those slices is determined, and this area is then multiplied by 1 mm (the thickness of the slices). The product of this volume and glomerular numerical density (from the physical disector) provides an unbiased estimate of total glomerular number.

5.3.2
Acid Maceration

It should be noted that the technique of acid maceration has also been used very successfully to count total glomerular number in kidneys. With this technique, kidneys are incubated with hydrochloric acid (for different lengths of time depending on the size of the kidney) and then shaken to dissociate renal components. When performed skilfully, a suspension of tubular structures and unbroken glomeruli is produced. Unfortunately, the results obtained with this approach of glomerular isolation and subsequent counting have not always been reliable, but several laboratories have used this technique with great skill and success (see for example Lelievre-Pegorier et al. 1998; Merlet-Benichou 1999; Gilbert and Merlet-Benichou 2000).

5.4
Microarrays

One approach to identifying genes that are expressed within the kidney at different developmental time points, under certain growth conditions or within various sub-compartments of the kidney is the use of DNA microarrays. This technology involves the isolation of mRNA from the sample of interest, which is converted into fluorescently labelled cDNA. This sample is hybridized to tens of thousands of cDNA/oligo clones that have been printed onto a glass slide. Microarrays allow researchers to analyse gene expression at a global level. A variety of commercially made mouse and human DNA microarray clone sets are now available. In addition, the use of bioinformatics to categorize the differentially expressed genes allows the researcher to select genes for further investigation based on gene ontology, biological pathways, and/or the membrane organization of the protein.

Several large-scale experiments have been conducted on the developing rat and mouse kidney, generating lists of temporally expressed genes (Stuart et al. 2003; Schwab et al. 2003; Challen et al. 2005; Martinez et al. 2006). However, in these studies, in order to determine the precise cellular localization of the gene expression in the kidney, RNA in situ hybridization or immunohistochemistry is required. Spatial gene expression studies have been designed to reduce the random chance of identifying a gene within a renal compartment of interest as in the temporal studies. The spatial gene expression experiments have investigated which genes are expressed in the metanephric mesenchyme vs ureteric epithelium (Stuart et al. 2003; Schwab et al. 2003; Takasato et al. 2004; Challen et al. 2005; Schmidt-Ott et al. 2005; Caruana et al. 2006a), uninduced MM vs induced MM (Valerius et al. 2002), tip ureteric epithelium vs trunk ureteric epithelium (Schmidt-Ott et al. 2005; Caruana et al. 2006a), glomeruli (Takemoto et al. 2006), side population (Challen et al. 2006) and genetically altered kidneys (Nishinakamura and Takasoto 2005; Schwab et al. 2006). To answer such questions, researchers have employed isolation techniques such as manual micro-dissection (Stuart et al. 2003; Schwab et al. 2003; Schmidt-Ott et al. 2005; Caruana et al. 2006a), fluorescence-activated cell sorting in conjunction with reporter mice in which the spatial compartment of interest is fluorescently tagged (Takasato et al. 2004; Challen et al. 2005) and magnetic Dynabeads (Takemoto et al. 2006). Recently, microarrays have been used to identify novel genes associated with congenital anomalies of the kidney in humans (Jain et al. 2007).

5.5
Use of Knock-in/Reporter Mice to Analyse Gene Function and Expression

As outlined in Sect. 3 above, much of our knowledge on the molecular regulation of kidney development has been derived from the analysis of genetically altered or knock-out mice. Gene inactivation in knock-out mice involves homologous recombination in which the gene of interest is excised and usually replaced by a reporter gene such as LacZ or GFP. The reporter is under the transcriptional control of the gene of interest and allows the researcher to visualize the expression pattern of the gene during development. These genetically altered mice that carry a reporter gene are often referred to as knock-in mice.

In many instances, gene inactivation of kidney expressed genes results in embryonic lethality prior to kidney development. To overcome this, researchers have devised strategies to produce conditional gene knock-outs. The most commonly used system involves Cre/loxP recombination. Cre recombinase, an enzyme produced by bacteriophage *P1*, is not present in mammalian cells. Cre mediates DNA recombination at 34-bp sequences, called loxP. To generate conditional knock-out/knock-in mice, one must generate a mouse line that harbours Cre-recombinase driven by the promoter of a tissue-specific gene. This mouse is then bred with a second mouse carrying a loxP flanked gene, the gene which one wishes to inactivate. When these mice are interbred, the tissue specific Cre recombinase will excise the gene flanked by the loxP. This methodology has resulted in the identification of several kidney-specific promoters that have been used to drive the expression of reporters such as lacZ or GFP.

Mice carrying reporter genes can be exploited to perform live cell imaging of the developing kidney. Fluorescent reporters allow live cell imaging to be performed on the developing metanephros in vitro, combining metanephric organ culture, confocal microscopy and time-lapse photomicroscopy to follow specific cell populations during organogenesis (as described above in Sects. 5.1 and 5.2). The best known reporter mouse used for studying renal development is the *Hoxb7/GFP* transgenic mouse developed by Srinivas et al. (1999a). As described in Sect. 5.1, in this mouse GFP is under the control of the *HoxB7* promoter, allowing visualization of the WD, ureteric tree and its derivatives (collecting ducts, calyces, renal pelvis, ureter). Images of kidneys of *Hoxb7/GFP* mice at different stages of embryonic development are shown in Fig. 9).

To visualize the progression of two or more kidney genes in combination, intercrosses of reporter mice carrying spectral-variant fluorescent reporters

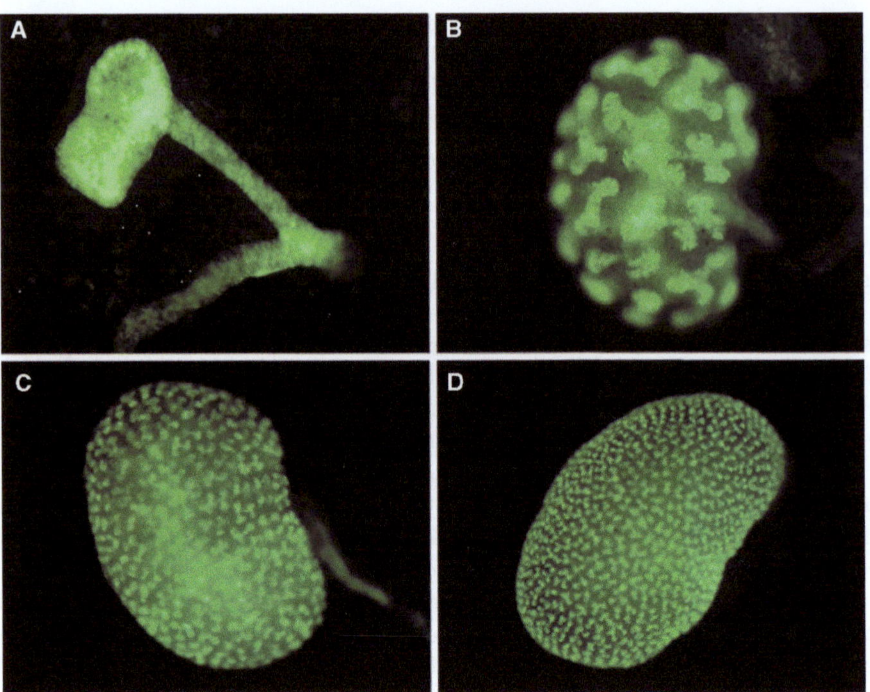

Fig. 9A–D Fluorescence photomicrographs of metanephroi from *Hoxb7/GFP* transgenic mice at **A** 11.5 dpc, **B** 13.5 dpc, **C** 15.5 dpc and **D** 17.5 dpc. Not to scale. (Images from Caruana et al. 2006 and reproduced with permission from *Nephron Experimental Nephron*, S. Karger and AG Basel)

(mRFLP, ECFP, EYFP) (Hadjantonakis et al. 2002, 2003; Long et al. 2005) can be undertaken. Reporter mouse lines can also be intercrossed with genetically altered mice to allow the researcher to follow the movement of a particular marked cell type in a mutant context. For example, the *Hoxb7/GFP* transgenic mouse line has been intercrossed with *Foxd1* null-mutant mice to analyse the effects of the loss of *Foxd1* on the branching pattern of the ureteric tree (Levinson et al. 2005). Some examples of reporter mice expressing GFP in embryonic kidney cell types include the *Hoxb7/GFP* transgenic mice, as mentioned above, *FoxD1* knock-in mice in which GFP marks stromal cells (Levinson et al. 2005), *NPHS1* transgenics in which GFP marks podocytes (Eremina et al. 2002), *Ksp-cadherin (Ksp1.3/BgEGFP)* transgenics in which GFP marks the UB and tubules (Shao et al. 2002), Tamm Horsfall protein transgenics in which GFP marks the thick ascending limb of the loop of Henle and early distal tubules (Zhu et al. 2002), and *Sall-1* knock-in mice in which GFP marks the mesenchyme, nephron structures and cortical stroma (Takasato et al. 2004).

6
Development of Function in the Fetus

6.1
Mesonephric Function

The mesonephros is the functional adult kidney in amphibia and fish and is an essential precursor for the formation of the metanephros. Thus, gene defects which prevent formation of a mesonephros inevitably result in total anephric neonates. The transient mesonephros consists, generally, of a small number of nephrons (30–70) which contain a glomerulus, proximal and distal tubules, but no loop of Henle (Moritz and Wintour 1999; Ludwig and Landman 2005; Wrobel 2001). In some species, the collecting duct drains via the cloaca and urachus, into the allantoic compartment. In those species in which the placenta is cotyledonary (sheep, cattle) or diffuse (pig), the accumulation of allantoic fluid allows the growth of the allantoic membrane to reach the whole uterine surface, including caruncles (the predetermined site of placenta formation) in both horns of the uterus. However, the allantois can also reabsorb fluid. Fetal vascularization of the placenta is carried via the allantois.

In many species, there is unequivocal evidence that the mesonephros functions both as an excretory organ (sheep, pig, rabbit) and as a source of circulating hormones and enzymes (erythropoietin, renin) and contains receptors for glucocorticoids, mineralocorticoids and growth factors such as insulin and IGF-1 (Peers et al. 2001; Kitraki et al. 1997; Leeson 1959; Korgun et al. 2003) as well as Ang II receptors (Butkus et al. 1997). In addition, the mesonephros has been shown to contribute cells to the developing gonads, and the aorta-gonad-mesonephric (AGM) region is a source of haemopoietic stem cells (Sainio and Raatikainen 1999). The glucocorticoid receptors found in the sheep were shown to be functional because a 48-h exposure of the developing fetus, at a time when only mesonephros was present (26–28 days), produced a significant alteration in allantoic fluid composition (Peers et al. 2001).

6.2
Fetal Metanephric Renal Function

The adult kidney has important functions in regulating the blood volume both by the regulation of plasma volume (salt and water) and the regulation of the haematocrit (red cell volume), as well as many factors which control blood pressure (Donnelly 2001). The newborn baby has a much higher percentage of water than the adult (Modi 2003) and the water content is even greater in premature babies (Omar et al. 1999), but skin permeability is higher, facilitating excess water loss. However, late hyponatraemia is very common in very low-birthweight infants (Roy et al. 1976). In utero, the major regulator of fluid balance is the placenta (Wintour 1998) and, indeed, one major function of the kidney is to secrete a dilute urine, thereby maintaining the volume of amniotic (and in some species, allantoic) fluid. A comparison of fetal and adult renal function is shown in Table 4. The fetal kidney achieves this

Table 4 A comparison between renal function in the fetus and adult[a]

Renal parameter	Fetus	Adult
Glomerular filtration rate (ml/kg/h)	60–70	120
Urine flow rate (ml/day/kg)	240	30
Renal blood flow (% of cardiac output)	3	25
Fractional sodium excretion (%)	3–5	<1
Free water clearance (ml/kg/day)	160	–90

[a] Most data for the fetus have been obtained from studies in the chronically cannulated ovine fetus over the last 50 days of pregnancy

aim even though it receives a comparatively small fraction of the cardiac output (3% vs 25% in the adult), because the mechanisms to retain sodium and water are much less mature in the kidney before birth. In fact, the normal urine production rate in a 2-kg ovine fetus is 0.5 l/day compared with 1 l/day in a 60-kg adult (Wintour 1998). The GFR in the fetus is actually quite high (1.8 l/kg body weight/day) relative to the low blood flow, and many agents (cortisol, Ang II) are diuretic and natriuretic in the fetus, but not in the adult, because they increase the filtered load beyond the capacity of the immature distal tube and collecting duct to reabsorb the fluid appropriately (Moritz et al. 2000; Towstoless et al. 1989). Drugs that are cleared by the kidney need to be given in much smaller doses in the neonate than suggested by body weight (Chen et al. 2006b). Metanephric glomerular filtration is thought to start at 14 dpc (mouse), 9–12 weeks (human) and 40 days (sheep).

In the fetal kidney, the fractional excretion of sodium is approximately 5%, compared to only 1% in the adult kidney, due to the incomplete maturation of ion transporters. The fetus also does not develop as high a concentration gradient as does the adult. In addition, the osmolality of fetal urine in the unstressed sheep/human fetus is always less than that of plasma, due at least in part to the low levels of aquaporin 2 gene expression in the collecting tubules (as outlined in Table 5 and discussed in Sect. 6.4). Normal urine osmolality is less than 150 mOsmol/kg water and maximal concentration capacity rarely exceeds 500 mOsmol/kg water. Thus, the newborn infant/lamb is somewhat comparable to an adult with nephrogenic diabetes insipidus.

6.3
Sodium Transporters

In Table 6 are listed the major sodium transporters of the adult kidney and their main mode of regulation; the ontogeny of each form, in so far as it is known, is also indicated (Knepper et al. 2003; Beutler et al. 2003; Holtback and Aperia 2003; Matsubara 2004). It is worth noting that the most-expressed isoform of several sodium channels (e.g. Na/K/ATPase-β; Na$^+$/H$^+$ exchanger) change after birth. In general, the total level of expression of sodium transporters is lower in the fetus than in the adult and this

Table 5 Location, ontogeny and regulation of aquaporins (AQPs) in the kidney

Aquaporin	Structural location	Ontogeny (age at which first expressed)	Age at which adult levels are achieved	Regulation
AQP1	Proximal tubule	16 dpc (rat)	>21 days (rat)	↑ angiotensin II, glucocorticoid
	Descending limb of loop of Henle	12/40 weeks (human)	15 Months (human)	
	Apical and basola-teral membranes	41/147 days (sheep)	6 Weeks (sheep)	
		All less than 50% by term		
	Arcuate artery and descending vasa recta	17 dpc (rat)	↓ After birth (rat)	
AQP2	Principal cells of collecting duct.	18 dpc (rat)	28 dpc (rat)	AVP
	Intracellular until translocation to apical membrane	100/147 days (sheep)	>42 days (sheep)	↑ Angiotensin II infusion (fetus)
		40% By term Low levels midgestation and <50% at term (human)		↓ Lithium Potassium deficiency, low protein diet
AQP3	Principal cells of collecting duct: basolateral membrane	18 dpc (rat)	Birth (rat)	↑ Aldosterone
AQP4	Principal cells of collecting duct; basolateral membrane	Not known	Not known	

accounts for the higher percentage of sodium excreted in fetal urine (5% vs 1%) even though fetal urine is normally hypotonic in the unstressed mammalian fetus (Wintour 1998). Evidence of the importance of each isoform can be seen from knock-out mice studies or congenital defects in the human. For example, in Bartter's syndrome (a mutation in the $Na^+/K^+/2Cl^-$ co-transporter), there is severe polyhydramnios (excess amniotic fluid), usually leading to premature delivery (Amsalem et al. 2003). It is also of note that epithelial sodium channels (ENaCs α, β and γ) are found in higher abundance in the adult female rat kidney than in the adult male; therefore, all experiments on programming using rats should consider sex of the animal (Gambling et al. 2004).

Table 6 The major sodium channels in the kidney and their regulation

Segment	Sodium transporter-fetal/adult	Regulation
Basolateral		
All	Sodium-potassium ATPase	↑ By sodium deficiency, aldosterone, angiotensin II, noradrenaline (α receptor)
	($Na^+/K^+/ATPase$ α1)	↓ By dopamine (D1)
	Rate-limiting	Atrial natriuretic factor
	Increased five- to tenfold postnatally	
	$Na^+/K^+/ATPase$ β1(2 before birth)	↓ By aldosterone, angiotensin II, sodium deficiency
	$Na^+/K^+/ATPase$ γ	MW of protein form changed by sodium deficiency
	Regulatory on activity	(85–70 kD)
Apical		
Proximal convoluted tubule	Sodium hydrogen exchanger (NHE3 in the adult, 1,2,4 before birth)	↑ By noradrenaline (a)
Proximal straight tubule	Major Na^+ transporter	Angiotensin II
		↓ By dopamine, parathyroid hormone
Thick ascending limb of loop of Henle: macula densa	$Na^+/K^+/2Cl-$ cotransporter	Inhibited by furosemide, bumetanide
	Defective in Bartter's syndrome	
	Present in fetal life (genetic defect →polyhydramnios)	
Distal tubule	$Na^+/Cl-$ cotransporter	Inhibited by thiazide
		Protein, but not mRNA.
		↑ By aldosterone, sodium deficiency
Cortical connecting tubule; collecting duct-cortical; outer medullary	ENaC α	Inhibited by amiloride
	Lower levels in fetus vs adult; higher in adult female than male	mRNA and protein

ENaC β	↑ By aldosterone, sodium deficiency, angiotensin II ↑ Relocation to membrane Protein but not mRNA ↓ by sodium deficiency, aldosterone, angiotensin II
Higher levels in fetus vs adult; higher in adult female than male	
ENaC γ	As for ENaCβ
Higher levels in fetus; higher in adult female than male	

6.4
Water Channels

In the adult kidney, a number of aquaporins (a generic name for water channels) are expressed, although the function of only four aquaporins (AQPs 1, 2, 3 and 4) have been studied in any detail. The ontogeny of AQP1 and AQP2 have been studied in rat, sheep and human, and the results from the human and sheep agree substantially. This is predominantly a result of nephrogenesis being completed before birth in these two species, whereas the rat kidney completes most (80% or more) nephrogenesis after birth. As seen in Table 5 the relatively poor ability of the fetal kidney to reabsorb water can be linked to the relatively low levels of AQP1 and AQP2 in the kidneys at birth. AQP1 is the major aquaporin of the proximal tubule and AQP2 is the only water channel expressed in the apical membrane of collecting duct cells. When water conservation is required, arginine vasopressin (AVP) is produced and acts on receptors on the basolateral membrane of the collecting duct's principal cells, to stimulate the phosphorylation and transport of vesicular AQP2 to the apical membrane, thus allowing water to pass through the cells in response to an osmotic gradient. AQP3 and 4 are located in the basolateral membrane to allow exit of reabsorbed water.

6.5
Renal Renin–Angiotensin System

An intact renal renin–angiotensin system (RAS) has long been known to be crucial for fetal renal development, as described in Sect. 3.5. All components of this system (renin, angiotensinogen, ACE and the angiotensin receptors) are present in the human fetal meso- and metanephros from very early in gestation allowing for local action (Schutz et al. 1996). The major action of the RAS during fetal development appears to be on the kidney where it acts to maintain GFR and ensure a high rate of urine production (Lumbers 1995). Recently, changes in the renal RAS were found in human fetuses and neonates with recessive tubular dysgenesis (RTD), which is characterized by the absence or poor development of proximal tubules and oligohydramnios (Lacoste et al. 2006).

7
Nephron Endowment

7.1
Nephron Endowment in the Human

It is important to emphasize that there is a wide range in nephron number in human subjects with apparently normal kidney morphology and renal function. This is contrary to what has commonly been reported in many textbooks, which state that the human kidney contains one million nephrons. Several studies in the past 20 years have estimated total nephron number in normal human kidneys using unbiased stereological methods (Nyengaard and Bendtsen 1992; Keller et al. 2003; Hoy et al. 2003; Hughson et al. 2003). It is reassuring that the mean values reported by these studies are relatively similar. Moreover, all studies have reported large ranges in nephron number, with the studies with the largest sample sizes reporting the largest ranges.

Nyengaard and Bendtsen (1992) used the physical disector/fractionator method to estimate (count) the number of glomeruli in an autopsy study of 37 adult Danish human subjects (16–87 years of age) with no evidence of renal disease. Total nephron number ranged fourfold, from 331,000 to 1,424,000, with a mean value of 617,000. In these apparently normal kidneys, Nyengaard and Bendtsen (1992) demonstrated an age-related decline in nephron number and a positive correlation of kidney weight with total glomerular volume but not number of glomeruli.

Subsequent to these studies, we have accumulated substantial stereological data of glomerular number and size in human autopsied kidneys. Our data is derived from female and male subjects of all ages that have undergone autopsy due to sudden or unexpected death (Hughson et al. 2003; Hoy et al. 2003). Importantly, in these studies we have examined kidneys of subjects from different ethnic origins: Aboriginal and Non-Aboriginal Australians, Senegalese Africans and non-African and African-Americans (descendants from countries such as Senegal). Hoy et al. (2003) reported nephron number in 78 kidneys from African-Americans and white Americans and from Aboriginal and white Australians. Subject ages ranged from newborns to 84 years. The American kidneys were collected at autopsy from Hinds County, Jackson, Mississippi, while the Australian kidneys were collected from the Top End of the Northern Territory. The main finding from this report was that total glomerular number ranged almost ninefold (from 210,332 to 1,825,380), with a mean value (SD) of 784,909 (314,686). Glomerular number was less in older subjects. This could be the result of age-related loss and/or due to older subjects being born with fewer nephrons. Females contained marginally fewer glomeruli than males and in this preliminary report no significant variations by ethnic group were observed. In a subsequent report, Douglas-Denton et al. (2006) provided an update on the nephron counts in more than 200 kidneys, including 19 Aboriginal Australians, 24 white Australians, 84 white Americans and 105 African-Americans. When classified according to race, mean nephron number in the Aboriginal Australians (713,209; range 364,161–1,129,233) was significantly less than in white Australians (861,541; range 380,517–1,493,665) ($p<0.05$). However, nephron number in American whites and blacks was similar (whites: 843,106, range, 227,327–1,660,232; African Americans: mean, 884938, range, 210,332–2,026,541). Recently, we have found that the total nephron number in 46 Senegalese Africans ranged 3.3-fold (536,171–1,764,421) with a mean of 992,353±271,582 in all subjects (B. McNamara et al., unpublished data). A trend towards declining nephron number with age was observed ($r^2 = 0.07$, $p=0.08$). Interestingly, nephron number in 38 Senegalese adults and 97 African-American adults differed by only 3.7% following adjustment for age and gender. Table 7 shows normal nephron number in the human in the reported studies in which reliable stereological methods have been used.

Table 7 Nephron endowment in humans

Nephron number	Range	Sample size	Sex (M/F)	Reference
617,000 (mean)	331,000–1,424,000	37	19 M/18 F	Nyengaard et al. 1992
1,429,200 (median)	~800,000–2,000,000	10	9 M/1 F	Keller et al. 2003
870,582 (mean)	227,327–2,026,541	208	Mixed	Hoy et al. 2005

These three studies have utilized optimal unbiased stereological methods (see text for details)

In these studies, nephron number has been significantly linked to gender (approximately 17% higher in men), age, race (lower in Australian Aborigines) and birthweight; glomerular number is also inversely correlated with glomerular size (Hoy et al. 2003; Hughson et al. 2003; Samuel et al. 2005). Taken together, our findings and that of the Danish group (Nyengaard and Bendsten 1992) demonstrate an age-related decline in nephron number. Indeed, it is conceivable that there is substantial loss of glomeruli due to acute and chronic insults throughout life. Therefore, the question arises: is the wide range in nephron endowment observed in the human population a consequence of postnatal hits on the kidney leading to variable loss of nephrons and thus directly impacting on nephron number? If this is the case, is the nephron compliment at birth relatively constant among all individuals? To date, there have been relatively few human studies that have quantified nephron endowment at birth. Indeed, there is compelling experimental evidence to demonstrate that insults in utero can directly impact on nephron endowment (see Sect. 8). However, there has only been one small human study (six appropriately grown stillbirths and six severely intrauterine growth-restricted [IUGR] stillbirths) that has looked at nephron number during normal fetal development in humans (Hinchliffe et al. 1992). In this study of stillborn infants, there was a significant correlation in nephron number with gestational age in appropriately grown fetuses, but in those infants that were severely growth-restricted, nephron number fell below the 10th percentile for gestational age, thus demonstrating a reduced nephron endowment in the IUGR subjects. In support of these findings, we have shown a direct correlation between birthweight and nephron number in human autopsied kidneys. For instance, in studies of Caucasian and African-Americans, there was a direct linear correlation between nephron number and birthweight; the linear regression coefficient predicted that there was an increase of approximately 250,000 nephrons per kilogram increase in birthweight (Hughson et al. 2003, 2006). However, it is to be noted that a high proportion of these kidneys were derived from adult human subjects. To further examine the relationship between birthweight and nephron number more closely, we have recently analysed nephron number in autopsied infant kidneys (<4 years of age); importantly in these 22 kidneys there is also a wide range in total nephron number from 210,332 to 1,391,375 (mean value of 697,824 nephrons) (J.F. Bertram, W.E. Hoy and M.D. Hughson, unpublished data). This finding strongly suggests that the wide range in nephron number observed in adult kidneys is present from birth. The gender-related differences in nephron endowment that we have observed may be linked to differences in birthweight between males and females (approximately 17% greater in males). Likewise, the reduction in nephron endowment observed in the Australian Aboriginal population when compared to non-Aboriginal Australians may be linked to the high incidence of low birthweight and/or prematurity in this population (Smith et al. 2000; Rousham and Gracey 2002).

7.2
Effect of Prematurity

Over the past two decades, the incidence and survival of infants after preterm birth has increased substantially (Noble 2003; Cockey 2004; Hoekstra et al. 2004). Despite significant improvements in the treatment of these preterm neonates, there remains a high incidence of renal failure in the neonatal period (ranging from 8% to 24%), which often leads to the death of these infants (Drukker and Guignard 2002; Andreoli 2004). Even with advanced management of acute renal failure in the preterm neonate, the mortality rate remains high (30%–60%) (Drukker and Guignard 2002; Andreoli 2004). Normally, postnatal renal adaptations in newborns include increases in GFR and sodium reabsorption, and there is evidence to suggest that this is delayed in preterm infants (Guillery 1997; Gallini et al. 2000; Awad et al. 2002). Thus, the immature preterm kidney may not be able to independently sustain renal function after birth. It is imperative to gain an understanding of the effects of preterm birth on the kidney. This is especially important since the formation of nephrons occurs predominantly during late gestation at a time when many preterm infants are already delivered (Hinchliffe et al. 1991). To date there is a striking paucity of information on the effects of preterm birth on the development of the kidney and this has been largely attributed to the lack of an adequate animal model to address this issue. In order to assess the effects of preterm delivery on renal development, it is imperative to use an animal model where the ontogeny of the kidney closely resembles that in the human and also where the postnatal care of the neonate is the same as that used in the human. We have recently reported the suitability of the baboon as a model to examine nephron endowment (Gubhaju and Black 2005). In this primate model, we have evidence to demonstrate that nephrogenesis continues after preterm birth (Gubhaju et al. 2005). Alarmingly, however, our preliminary findings suggest that the glomeruli formed in the extrauterine environment may be abnormal.

A recent histomorphometric analysis of autopsied kidneys from deceased preterm infants suggests that the absolute number of nephrons is reduced in the preterm infant, with the number of medullary ray generations of glomeruli reported to be less in preterm kidneys than in term kidneys (Rodriguez et al. 2005).

A reduced number of glomerular generations and glomerulomegaly has also recently been reported in the kidneys of a 10-year-old child who was born preterm and subsequently developed renal failure (Rodriguez et al. 2005). In another recent follow-up study of young adult subjects that were born very premature, it was those that were IUGR at the time of delivery that exhibited the most severe renal dysfunction, and the authors suggest that the findings may be linked to impaired nephrogenesis in these subjects (Keijzer-Veen et al. 2005). However, it should be emphasized that interpretation of the limited human autopsy data available to us at the present time must be treated with caution, since it is confounded by the fact that the kidneys analysed were from deceased preterm infants and so effects in the kidney may be due to the failure to thrive after birth rather than preterm birth per se.

Thus, we suggest the preterm infant that is small for gestational age at delivery is likely to be most at risk of reduced nephron endowment and kidney function impairment. Importantly in this regard, in a recent study of 172 preterm infants it was the very-low-birthweight infants who had the highest risk of acute renal failure (79% of cases were <1500 g) (Cataldi et al. 2005).

7.3
Factors Important for Normal Kidney Development

7.3.1
Vitamin A

Micronutrients such as vitamin A and iron, as well as adequate protein, calories and oxygen, are critical for normal renal development. Vitamin A deficiency (less than 0.7 mmol/l serum retinol), sufficient to cause night-blindness, affects up to half the pregnant women in Nepal (Haskell et al. 2005) and has been estimated to affect at least 20 million pregnant women globally (Christian 2003). Retinol (vitamin A) is converted to retinoic acid and acts on a dimeric intracellular receptor (RAR+RXR) to influence many developmental processes (Wei 2004). Mild vitamin A deficiency has been associated with a decreased nephron number in experimental animals (Lelievre-Pegorier et al. 1998; Burrow 2000). Whilst vitamin A deficiency affects the development of many organs, including the placenta, the expression of two genes shown to be critical for kidney development have been shown to be downregulated: these are the c-Ret receptor (Batourina et al. 2001) and midkine, a heparin-binding growth/differentiation factor (Vilar et al. 2002).

7.3.2
Iron

Iron deficiency during pregnancy is relatively common. It has been reported that true anaemia, with haemoglobin levels of 80–100 g/l, is seen in half of all pregnant women in the world (van den Broek 2003). The effect of iron deficiency has been studied in pregnant rats (Crowe et al. 1995; Lisle et al. 2003) and seen to result in the development of hypertension in the offspring, which was accompanied by a decrease in nephron number in the female offspring at least. There is some evidence that this may be related to an impairment of retinol mobilization from the liver (Strube et al. 2002) and thus micronutrient deficiencies (vitamin A and iron) might coexist and synergize in exacerbating the problem.

7.4
Timing of Nephrogenesis and Nephron Endowment in Animal Models

A large number of animal species are utilized to study development of the kidney and therefore it is critical to appreciate the differences between species, both in terms of the timing of nephrogenesis and the final number of nephrons formed. The most widely used animals in renal research are rodents (rats and mice), with mice in particular being used for genetic/molecular studies. One of the major differences in these models compared to the human is the timing of nephrogenesis and the number (and range) of nephrons formed. In both the rat and mouse, development of the permanent metanephric kidney begins soon after mid-gestation and is not completed until 1 week or more postnatally. As noted in Sect. 2.3, the human kidney has completed nephrogenesis by week 36 of gestation, although nephron growth and differentiation continues for some time after birth. This means that nephron endowment in the human, at least in most cases, is dependent totally upon the in utero conditions and is not influenced by the postnatal environment, whilst in the rodent there is a significant period after birth where external factors may play a role. A list of species in which the timing of nephrogenesis is known appears in Tables 8, 9 and 10. For clarity we have expressed the period of nephrogenesis as a percentage of gestation to highlight the large variation.

It is also of great importance to consider the range of nephron number seen in the laboratory animal. Table 10 shows estimated glomerular (and thus nephron) number in some common experimental animals. Where possible, we have only included studies in which unbiased stereological methods were used. As can be seen, there is generally a very tight range of nephron number in rodent strains. The standard

Table 8 Timing of mesonephric development (as a percentage of gestation) in a number of animal species[a]

Species	Mesonephros present	Overlap with metanephric development
Human	4–16 weeks	11 weeks
Sheep	17–57 days	27 days
Rat	12–17 days	5 days
Mouse	10–13 days	2 days

[a]In many species (particularly those of long gestation) there is a period of considerable overlap where the mesonephros and metanephros co-exist. Of particular relevance in Table 9 is the fact that some animals complete nephrogenesis prior to birth while in others this process continues for a time postnatally

Table 9 Timing of metanephrogenesis (as a percentage of gestation) in a number of animal species[a]

Species	Approximate period of nephrogenesis	Approximate length of pregnancy	Relative period of gestation
Human	5–36 weeks	40 weeks	12–90%
Sheep	30–130 days	150 days	20–90%
Guinea pig	22–55 days	63 days	35–90%
Spiny mouse	19–37 dpc	40 days	50–90%
Mouse	11 dpc-PN 5–7 days	20 days	55–125%
Rat	12 dpc-PN 8–10	22 days	55–140%
Rabbit	12 dpc-PN21	32 days	35–160%
Pig	20 dpc-PN 21–25	112 days	20–120%

[a]In many species (particularly those of long gestation) there is a period of considerable overlap where the mesonephros and metanephros co-exist. Of particular relevance is the fact that some animals complete nephrogenesis prior to birth while in others this process continues for a time postnatally

Table 10 Total nephron number (in one kidney) in some commonly used experimental animals[a]

Species/strain	Nephron number	Sex	Age examined	Reference
Rat				
Sprague-Dawley	31,764 ± 3,667	M	Adult	Bertram et al. 1992
Sprague-Dawley	27,208 ± 1,534	M	22 weeks	Woods et al. 2004
Sprague-Dawley	26,248 ± 1,292	F	22 weeks	Woods et al. 2005
Sprague-Dawley	24,866 ± 1,261	F	30 days	Singh et al. 2007a
	28,713 ± 1,681	M		
Wistar-Kyoto	27,191 ± 3,512	M	4 weeks	Zimanyi et al. 2002
Milan-Normotensive	28,050 ± 561	Not reported	6–9 months	Menini et al. 2004
Spontaneously hypertensine (SHR)	28,620 ± 1,643	M	4 weeks	Black et al. 2002
Mouse				
C57/Bl6	11,886 ± 1,277	M	30 days	Cullen-McEwen et al. 2001
C57/Bl6	10,695 ± 864	F	30 days	Hoppe et al. 2007a
	10,755 ± 937	M		
Spiny mouse	7,245 ± 280	M	10 weeks	Dickinson et al. 2005
Sheep				
Merino	365,672 ± 36,016	M	130 days (fetal)	Douglas-Denton et al. 2002
Merino	402,787 ± 30,458	F	7 years	Wintour et al. 2003
Border-Leicester/ Merino cross	559,000 ± 198,000	M/F (singles)	140 days	Mitchell et al. 2005
	343,000 ± 106,000	M/F (twins)		

[a]All these studies utilized unbiased stereology to obtain nephron number

deviation in any one control cohort from one laboratory is approximately ±20% of the mean. This is not surprising given the years of inbreeding of laboratory animals, but it raises the question as to the relevance to the human. For example, if a given treatment reduces nephron endowment by 20% in a group of rats, this would be highly statistically significant. However, a 20% decrease from the mean in a cohort of humans would give values still well within the normal range. Another confounding factor is the effect of litter size on nephron endowment, since birthweight itself is largely dependent upon litter size. Animal models, particularly rodents, do, however, allow researchers to control for factors such as age at which nephron number is determined, sex and strain differences, and the effect of prenatal insults.

The sheep has been used as a unique animal model in which it is possible to study fetal renal function in utero because it is possible to chronically cannulate the fetus over the second half of pregnancy. This model has the advantage that nephrogenesis is complete about 3 weeks prior to birth, making it similar to the human. Nephron number in this model has been estimated in relatively few studies, but numbers still appear to fall over a much smaller range (up to twofold) than in humans (Douglas-Denton et al. 2002; Wintour et al. 2003; Mitchell et al. 2004; Zohdi et al. 2007). Nephron endowment in different strains of sheep appears to differ with desert adapted breeds, such as the Merino, having fewer nephrons than other breeds. Interestingly, twin lambs were found to have significantly fewer nephrons than singleton lambs (Mitchell et al. 2004).

8
Developmental Programming of the Kidney

8.1
Overview: The Developmental Origins of Health and Disease Hypothesis

As noted in the introduction, renewed interest in renal development, and in particular in the formation of an adequate nephron endowment, has occurred over the last decade, with worldwide interest in the concept that many diseases with onset in adult life in fact have their origins during development. During the 1990s, this concept was known as the Barker hypothesis or the fetal origins of adult disease, but subsequent studies demonstrating the crucial role played by the early postnatal environment have led to the developmental origins of health and disease (DOHaD) hypothesis. Simplistically, it is proposed that upon exposure to an insult or sub-optimal exposure in utero, the fetus makes adaptations to ensure short-term survival; however, many of these adaptations may increase the subsequent risk of developing particular diseases in adulthood. Low birthweight has been taken as a marker of a poor intrauterine environment. Of course, in reality, the situation is much more complex, with the eventual disease outcome being highly dependent upon interactions with the environment including lifestyle choices. It is not within the scope of this review to discuss in detail the extensive epidemiological and experimental studies undertaken to test this hypothesis; however, an understanding of the adult disease outcomes is crucial to fully appreciate the fundamental role that altered kidney development may play. In the remaining subsections of Sect. 8, we shall consider the evidence that altered renal development is a common underlying mechanism through which many prenatal perturbations may result in adult disease and explore the potential ways in which renal development may be affected. Finally, although the impact of developmental programming in the aetiology of many diseases is unknown, it is likely to be very significant. For example, it has been estimated that if all individuals were in the most favourable tertile of body size at birth and remained so throughout childhood, the incidence of coronary artery disease would be reduced by 40% in men and 63% in women (Barker et al. 2002).

8.2
Human Epidemiology

8.2.1
Links of Disease Outcome to Birthweight

David Barker, an English epidemiologist, demonstrated that the highest rates of infant death in Britain in the early 1900s coincided geographically with areas which had the highest incidence of death due to ischaemic heart disease and coronary artery disease many years later (Barker and Osmond 1986; Barker 1998). Barker and colleagues postulated that the high infant mortality reflected high levels of growth restriction in utero and suggested that poor fetal growth was linked to the subsequent development of cardiovascular disease in adulthood. They went on to demonstrate in a cohort of more than 15,000 people that those whose birthweight was between 8.5 and 9.5 pounds had approximately half the risk of dying from coronary heart disease as those born weighing less than 5.5 pounds (Osmond et al. 1993). Numerous subsequent studies over the last 10–15 years from around the world now broadly support this hypothesis (Godfrey 2006), not only for coronary heart disease but also many other forms of cardiovascular disease including stroke (Martyn et al. 1996) and hypertension (Curhan et al. 1996). It must be remembered in these studies that low birthweight refers to babies born at term (small for dates) rather than preterm or premature (see Sect. 7.2). In addition, although low birthweight is used as an indicator of poor fetal growth, often little information is available as to when and why this occurred. As detailed in Sect. 8.3.3, in many animal models, particularly those employing short-term maternal glucocorticoids as the prenatal insult, developmental abnormalities and adult disease can result without discernible changes in fetal growth. Nevertheless, particularly in human studies, birthweight is easily obtained and in clinical practice remains a surrogate marker for a suboptimal intrauterine environment.

Metabolic disease (including impaired glucose tolerance and non-insulin-dependent diabetes mellitus) is also strongly linked to a low birthweight (Hales and Ozanne 2003; Curhan et al. 1996). More recently other diseases, including chronic obstructive airway disease (Barker 1998; Lucas et al. 2004), osteoporosis (Harvey and Cooper 2004) and even mental disorders such as schizophrenia (Susser et al. 1996), have been linked to low birthweight.

Finally, although the data is not yet as strong, there is growing evidence that low birthweight is a risk factor for renal disease (Lackland et al. 2000; Hoy et al. 1999). This is of huge importance as in all Westernized countries the rates of chronic renal disease are increasing. Low birthweight has been identified as a progression promoter for renal disease (Alebiosu 2003).

8.2.2
Importance of Early Postnatal Growth

Growth during the first year of life has emerged as an independent risk factor for cardiovascular and metabolic disease. In fact, weight at 1 year is considered a better predictor of hypertension than birthweight (Eriksson et al. 1999). This suggests that the early postnatal environment can modify (either accentuate or attenuate) the effects of a poor intrauterine exposure. The full effects of catch-up growth or, conversely, a failure to thrive during early postnatal life have yet to be fully elucidated. In terms of the kidney, nutrition during this period in the human born at term cannot influence nephron formation, but this may be a period when compensatory changes in response to a low nephron endowment begin to occur and thus it represents a period where intervention/preventative measures may be possible. Nutrition in adulthood, especially intake of high salt and/or high-fat diets may be a compounding factor upon prenatal and early postnatal influences. This is evident in recent epidemiological studies from developing countries, such as India, which strongly suggest that being born at a low birthweight, experiencing poor growth in infancy (often due to infections and weaning practices) and becoming obese in adulthood is resulting in epidemics of coronary heart disease and type 2 diabetes (Fall 2001; Fall and Sachdev 2006).

8.3
Common Animal Models Used to Test the DOHaD Hypothesis

8.3.1
Maternal Undernutrition

The vast majority of experimental studies set up to test the DOHaD hypothesis have utilized some form of undernutrition. Depriving the mother (generally rats but sheep, mice and guinea-pigs have been used) of calories (global undernutrition) or specific components of the diet (most often protein deficiency but vitamin A and iron deficiencies have also been employed) for all of or specific parts of pregnancy can result in IUGR and offspring of a low birthweight. Whilst the protein-deficiency model undoubtedly results in a reduction in offspring birthweight, the long-term outcomes are highly variable and largely depend upon the specific dietary composition. A review by Armitage et al. (2005) highlights that two diets containing the same amount of protein often derive the calories from vastly different sources (either sugar or fats), and these differences may be just as important as the low protein per se. In many cases, however, the low protein results in offspring with elevated blood pressure in adulthood, lending support to the DOHaD hypothesis (Woods et al. 2004; Langley-Evans et al. 1999). A common and justified criticism of the undernutrition model is its relevance to women and their babies in Western society. Often the level of deprivation required to produce consistent physiological outcomes is quite extreme (5% or 8% protein instead of 20% in isocaloric diets, or a reduction to <20% of normal food intake) (Woods et al. 2001; Bloomfield et al. 2003). Whilst it is acknowledged that certain racial and socio-economic groups within first-world countries may experience calorie or protein restriction during pregnancy, other prenatal insults or exposures are likely to be of greater importance. This does not detract from the relevance of this model to developing countries where nutrition during pregnancy may be very poor and IUGR is exceedingly common (De Onis et al. 1998).

8.3.2
Placental Insufficiency

In Western society, the majority of pregnancies resulting in IUGR are thought to result from placental insufficiency. To model this in the rat, bilateral uterine vessel ligation has been performed at approximately day 16–18 of gestation (out of a 22-day pregnancy). This results in an abrupt loss of oxygen and nutrients to the fetus and causes significant growth restriction (Wlodek et al. 2005; Lane et al. 1998). It is also associated with increased pup mortality in utero and altered growth postnatally due to effects on maternal lactation (Wlodek et al. 2007). In the sheep, placental restriction has been modelled by either removing sites of placentation prior to pregnancy or infusion of microspheres into the circulation during the last third of pregnancy (Robinson et al. 1979; Murotsuki et al. 1997). Both regimes are capable of significantly inhibiting fetal growth. In the rat, there is evidence that placental insufficiency results in elevated blood pressure in male offspring (Wlodek et al. 2007).

8.3.3
Maternal Glucocorticoid Exposure

The third model of importance in the DOHaD field, one which has been used extensively by the authors, is that of short-term maternal glucocorticoid exposure. Studies in the sheep and rat have shown that maternal exposure for just 2 days during gestation (at certain critical windows) either to synthetic glucocorticoids (dexamethasone or betamethasone) or the naturally occurring one (cortisol or corticosterone, depending on the species), results in offspring with hypertension (Dodic et al. 1998, 2002; Ortiz et al. 2003; Singh et al. 2007a). This model differs somewhat from the undernutrition and placental insufficiency models in that there are no discernible changes in birthweight, suggesting IUGR is not a prerequisite for disease onset. An important aspect of this model is that the timing of exposure is of utmost importance, with exposures at certain developmental periods resulting in hypertension but not others (Dodic et al. 1998; Ortiz et al. 2003). As discussed in Sect. 8.4.1, the most sensitive time appears to be very early in renal development, irrespective of the actual stage of pregnancy.

8.4
Reduced Nephron Endowment: A Common Denominator in the DOHaD Hypothesis?

The wide range of models used to explore and test the DOHaD hypothesis has resulted in a large number of organs and systems being implicated as potential mechanisms leading to disease development. Although many changes are model- or species-specific, a striking observation has been the finding that a reduced nephron endowment is found following a wide range of insults (including those discussed in Sect. 8.3) in a number of species (rat, sheep, mouse). As noted elsewhere in this review, determination of nephron endowment is best performed using unbiased stereological methods because other methods are inherently biased and unreliable. The evidence discussed in the following sections therefore only includes studies in which unbiased stereology has been used (unless otherwise stated).

8.4.1
Experimental Manipulations Resulting in a Reduced Nephron Endowment

A summary of the experimental manipulations shown to affect nephron endowment is shown in Table 11. As can be seen, maternal undernutrition (global, low protein, iron deficiency) consistently results in offspring with a reduced number of nephrons, although this does not always result in elevated blood pressure.

Perhaps though, the strongest evidence demonstrating the importance of nephron endowment comes from the maternal glucocorticoid exposure model which, as noted in Sect. 8.3.3, is highly dependent upon the timing of exposure. When performed between 26 and 28 days gestation in the sheep (gestation=150 days), 13.5/14.5 dpc in the rat (gestation=22 days) or 20–23 dpc in the Spiny mouse (gestation=40 days), offspring are found to have an approximate 30% reduction in nephron number (Wintour et al. 2003; Singh et al. 2007; Dickinson et al. 2007). As can be seen, the relative stage of gestation ranges from very early in pregnancy (sheep), through to mid-pregnancy (Spiny mouse) to mid-late pregnancy (rat). However, renal development in the three species at the time of the glucocorticoid exposure is similar in all species with the UB having just invaded the MM and begun branching (Fig. 10). This strongly suggests that early renal development is particularly susceptible to insult.

Very recently, however, the crucial importance of the latter stages of nephrogenesis has also been demonstrated. In a rat model of uteroplacental insufficiency, pups born growth-restricted and fed by a mother with impaired lactation postnatally have reduced nephron endowment (Wlodek et al. 2007). However, cross-fostering of a pup born growth-restricted onto a mother with improved lactation is

Table 11 Experimental manipulations known to affect total nephron number[a]

Species	Time in development	Insult	Reference
Undernutrition			
Rat	Throughout pregnancy	Maternal low protein	Woods et al. 2004 2005; Hoppe et al. 2007b
Rat	Throughout pregnancy and up to postnatal day 10	Maternal low protein	Zimanyi et al. 2002
Rat	Postnatal day 1 until weaning	Postnatal food restriction	Schreuder et al. 2006
Sheep[b]	28–80 days of gestation	Global food restriction	Gopalakrishnan et al. 2005
Mouse	Throughout pregnancy and postnatal life (day 30)	Maternal low protein	Hoppe et al. 2007a
Maternal glucocorticoid exposure			
Sheep	Day 26–28 of gestation	Maternal dexamethasone	Wintour et al. 2003
Rat	14.5/15.5 dpc	Maternal corticosterone	Singh et al. 2007a
Spiny mouse	20–23 dpc	Maternal dexamethasone	Dickinson et al. 2007
Placental insufficiency			
Rat	18 dpc to term	Uterine vessel ligation	Wlodek et al. 2007
Rat	17 dpc to term	Uterine vessel ligation	Schreuder et al. 2005
Sheep	100–130 days of gestation	Placental embolization	Zodhi et al. 2007
Other			
Pig	PN2	Unilateral ureteropelvic obstruction	Eskild-Jensen et al. 2002
Sheep[c]	Day 100 of gestation	Unilateral nephrectomy	Douglas-Denton et al. 2002

[a]Studies have utilized unbiased stereology to determine nephron endowment except as indicated
[b]Indicates acid maceration. The insult produces a decrease in nephron endowment except as indicated
[c]Indicates an increase in nephron number

able to restore nephron endowment and prevent the onset of hypertension in male offspring (Wlodek et al. 2007). In a different model, postnatal growth restriction induced by placing pups of normal birthweight onto a dam, so that each pup has a reduced milk intake, also results in a reduction in nephron endowment (Schreuder et al. 2006). In the sheep, placental embolization beginning around day 100 of gestation can also reduce nephron number (Zohdi et al. 2007). These studies highlight that there is probably a number of critical windows when a perturbation of the in utero environment can influence nephron endowment.

A Rat metanephros at E14 (22 day gestation)

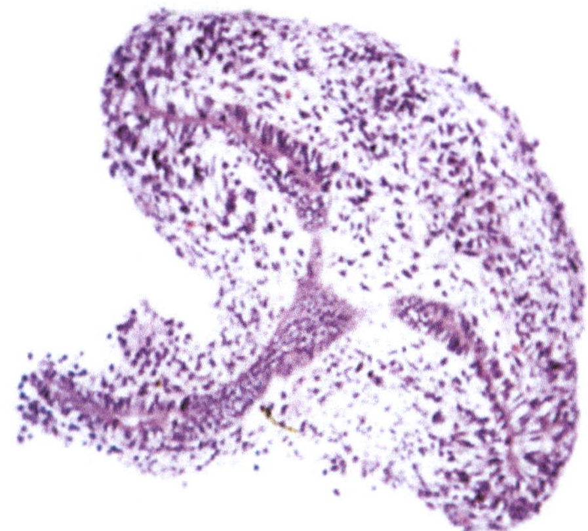

B Sheep metaneprhos at 27 days (150 day gestation)

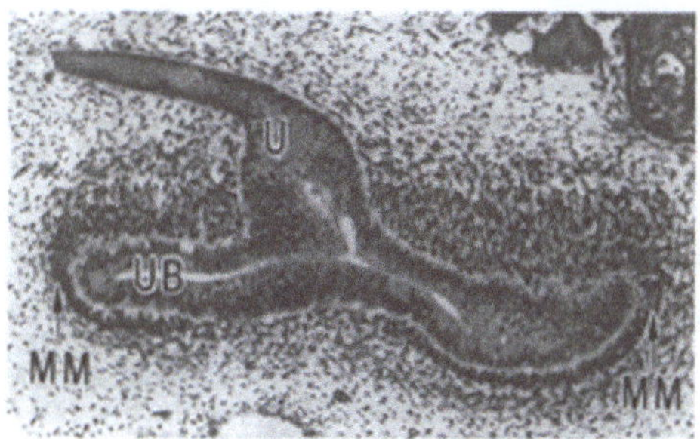

C Spiny mouse at 19 days (39-40 day gestation)

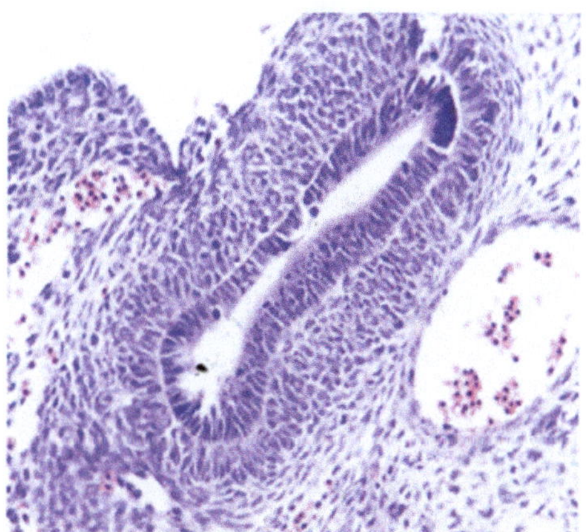

Fig. 10A–C The kidney of **A** rat at 14 dpc, **B** sheep at day 27 and **C** Spiny mouse at 19 dpc of pregnancy. Maternal glucocorticoid exposure at these times in these species can result in a nephron deficit in the adult. Note the relative similarity in the stage of development; there is a prominent ureteric bud but little, if any glomerular formation

8.5
Link Between Nephron Endowment and Hypertension

8.5.1
The Brenner Hypothesis

It is nearly 20 years since Barry Brenner and colleagues first proposed that "the renal abnormality that contributes to essential hypertension is a reduced number of glomeruli and tubules, the consequences of which are limitations in the ability to excrete sodium" (Brenner et al. 1988; Brenner and Anderson 1992). Brenner and colleagues proposed that a congenital nephron deficit was likely to be associated with increased hydrostatic pressure in the glomerular capillaries and glomerular hyperfiltration which would result in glomerular hypertrophy. This would act to maintain overall normal GFR, but this adaptation could only suffice for a defined period and eventually glomerulosclerosis would result, further perpetuating nephron loss. This is represented schematically in Fig. 11.

8.5.2
The Link Between Low Nephron Number and Hypertension in Humans

Very few studies have examined the relationship between nephron number and hypertension in humans. Indeed, we are aware of only three studies to date. In the first report, Keller et al. (2003) found that the kidneys of ten subjects with a history

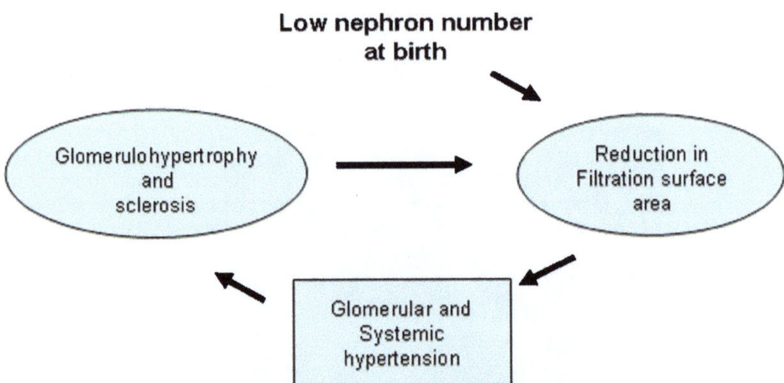

Fig. 11 The basis of the original Brenner hypothesis through which congenital nephron deficit (reduced nephron endowment at birth) results in hypertension. Note this is a loop with hypertension contributing to further glomerular loss, which in turn impacts on the high blood pressure. Current research suggests that a nephron deficit alone may not result in hypertension but in conjunction with compensatory changes within the kidney and other secondary insults (such as outlined in Fig. 10); hypertension may result (see text for details)

of hypertension contained approximately half the number of nephrons (median 702,379) as ten age-, sex-, height- and weight-matched subjects with no history of hypertension (median nephron number 1,429,200). In our recent study of human nephron number, we found that the kidneys of Australian Aborigines contained approximately 30% fewer nephrons than non-Aborigines (Hoy et al. 2006). The kidneys of Aborigines with a history of hypertension contained approximately 30% fewer nephrons than Aborigines without such a history ($p = 0.03$).

In a study of 62 African Americans and 84 white Americans, we found that African Americans with a history of hypertension had the same number of nephrons as those with no history of hypertension (867,359 vs 961,840; $p = 0.285$) (Hughson et al. 2006). In contrast, the kidneys of hypertensive white Americans contained fewer nephrons than the kidneys of normotensive whites (754,319 vs 923,377; $p=0.03$). Taken together, the data indicate that low nephron number and possibly low birthweight may play a role in the development of hypertension in white Americans and Australian Aborigines but not African-Americans.

8.6
Mechanisms Leading to a Reduced Nephron Endowment

Given that nephron number appears to be important in the programming of adult disease, it is worth considering how the various perturbations during development may influence the process of nephron formation. We have suggested that two likely processes resulting in a reduced nephron endowment are a decrease in branching morphogenesis and/or increased apoptosis (Moritz and Bertram 2006). Recently, we tested the involvement of branching morphogenesis using in vitro organ culture. Metanephroi from 13.5 dpc rats were grown in the presence of varying concentrations of dexamethasone and the degree of branching morphogenesis assessed by quantifying ureteric branch points. We found that addition of dexamethasone to culture media for 2 days was able to decrease the number of ureteric branch points. Even after the dexamethasone was removed and the kidneys allowed to grow in culture for a further 3 days, nephron number was significantly reduced (Singh et al. 2007b). We demonstrated that the expression of key genes involved in the regulation of branching morphogenesis (such as *GDNF*, *TGF-β1* and *BMP-4*) was altered both in vitro as well as in the embryonic kidney following maternal exposure to dexamethasone in vivo (Singh et al. 2007b). These studies are the first to clearly demonstrate a role for altered branching morphogenesis in a model of reduced nephron endowment.

Apoptosis (programmed cell death) is a normal part of kidney development. Alterations in expression of genes involved in apoptosis were identified in experiments designed to identify renal gene changes following maternal exposures to low protein (Welham et al. 2002). Subsequent experiments confirmed increased apoptosis in the kidneys of late gestation rats from dams on a low-protein diet (Welham et al. 2005). We have also found changes in apoptotic genes following glucocorticoid exposure in the Spiny mouse (Dickinson et al. 2007).

8.7
Programmed Changes in Renal Gene Expression

In addition to reducing nephron endowment, many maternal perturbations cause both short- and long-term changes in renal gene expression. Many of the gene changes that occur during development may contribute to the low nephron number. However, other changes, particularly in genes that regulate renal function in the adult, are likely to play a role in altering renal function. These changes are likely to underpin the onset of disease development.

8.7.1
Renal Renin–Angiotensin System

The renal RAS has been identified as a hormonal system that is altered in models of developmental programming both during development where alterations are likely to affect development and in the adult where function may be affected. During the period of nephrogenesis, the renal RAS is inhibited in offspring of rat dams exposed to protein restriction during pregnancy (Woods et al. 2001) and expression of the angiotensin type 1 (AT1) receptor is reduced (Vehaskari et al. 2004). Maternal corticosterone or dexamethasone exposure in the rat also alters AT1 receptor expression in the fetal rat kidney (Singh et al. 2007a, 2007b). Ang II binding to the AT1 receptor has been shown at least in culture systems to mediate the process of branching morphogenesis (Yosypiv and El-Dahr 2005). Thus, decreased expression of the AT1 receptor may result in less branching of the UB and thus a reduction in nephron endowment. Alterations in the expression of the AT2 receptor has been reported in the fetal kidney following glucocorticoid exposure in the rat (Singh et al. 2007a) where it may play a role in apoptosis (Tufro-McReddie et al. 1995). Interestingly, changes in the RAS appear to be gender-specific, which may contribute to the differential outcomes seen in males and females (discussed in Sect. 8.8.1) (McMullen and Langley-Evans 2005). After completion of nephrogenesis, components of the RAS, particularly the AT1 receptor, appear to be upregulated in the kidneys of offspring that have been programmed in utero. This may provide a mechanistic link to the onset of hypertension as there is now evidence to suggest that increased activation of the renal RAS may cause sodium retention and contribute to sustained elevations in blood pressure (Ichihara et al. 2004). Rats and sheep exposed to glucocorticoids (Singh et al. 2007a; Moritz et al. 2002a), rats exposed to low protein (Sahajpal and Ashton 2003) and male rats exposed to placental insufficiency (Wlodek et al. 2007) all have increased gene expression of one or more of the AT1 receptors.

8.7.2
Sodium Channels

There have been three reports of programmed changes in renal sodium transporters. Bertram et al. (2001) reported that the kidneys of 12-week-old offspring of rats, which had been protein-deprived (9% vs 18%) during pregnancy, had higher levels of mRNA for both α_1 and β_1 subunits of $Na^+/K^+/ATPase$ than controls. This may signify that the tubules were more adult-like in these offspring; when kidneys are stimulated to increase sodium retention by aldosterone or Ang II in adult animals they normally increase expression of the α subunit of $Na^+/K^+/ATPase$ but decrease the β subunit (Knepper et al. 2003; Beutler et al. 2003). In the second study, however, both the $Na^+/K^+/2Cl^-$ transporter and Na^+/Cl^- co-transporter were upregulated in the 4-week old offspring of rats fed a lower protein intake (6%) from day 12 to parturition, without changes in Na^+/H^+ exchanger isoform 3(NHE3) or any ENaC channels (Manning et al. 2002). In a subsequent study (Vehaskari et al. 2004), it was shown that similar treatment upregulated the AT1a receptor at 4 weeks, although it had been lower than normal in the newborn kidney. Treatment of pregnant rats with dexamethasone from days 15–18 of pregnancy resulted in hypertension in the offspring at 7–8 weeks, with evidence of increased protein levels of NHE3 at the brush-border membranes of proximal tubule cells, and a 50% increase in in vitro activity of this transporter, though mRNA values were unchanged (Dagan et al. 2007).

8.8
Why a Low Nephron Endowment Does Not Always Lead to Hypertension

In the previous section, we have considered the evidence suggesting that a low nephron endowment is an important factor in the development of hypertension in the adult. However, a large (and growing) body of evidence suggests low nephron endowment alone does not always result in elevated blood pressure (Zimanyi et al. 2004; for review see Moritz and Bertram 2006). This indicates that a low nephron number is unlikely to be the sole factor contributing to elevated adult blood pressure and has caused some sceptics to totally dismiss the importance of low nephron endowment. However, over the last 5 years, we have stressed that it is a low nephron number in combination with other (detrimental) compensatory changes that are crucial to the development of disease (Moritz et al. 2003). After a prenatal insult, there are likely to be many compensatory changes in the kidney. These compensatory changes are likely to occur in stages. Whilst the initial changes may be beneficial to maintain renal function (e.g. glomerular hypertrophy), secondary changes (e.g. glomerulo-sclerosis) may lead to renal damage and perpetuate disease. Many factors have the potential to modify the long-term outcome and knowledge of the interaction between these factors, and the low nephron environment will aid our overall understanding of the underlying mechanisms leading to disease outcome. In Fig. 12 we highlight some of the factors that may affect the overall disease outcome.

8.8.1
Effect of Gender/Sex

Sex differences in the control of renal and cardiovascular physiology are becoming abundantly clear (Denton and Bayliss 2007). This is also the case in the field of DOHaD. Male offspring subject to a particular maternal perturbation are nearly

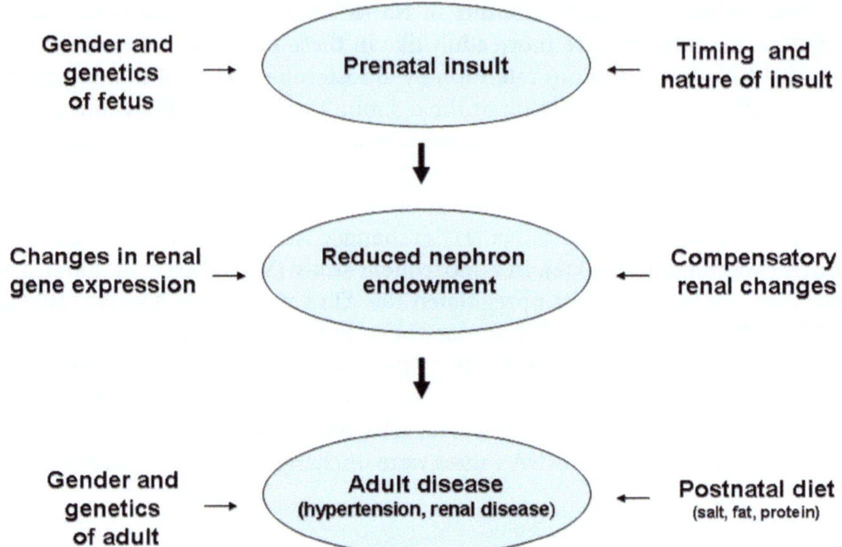

Fig. 12 Factors which may impact on the final disease outcome following a prenatal insult with a particular focus on the role of the kidney. It should be noted that other organs and systems may interact with renal changes to exacerbate or ameliorate the onset of disease

always more affected than females. Woods et al. (2001, 2005) showed that moderate protein restriction in the rat caused a nephron deficit in both sexes, but only males developed hypertension; it was only with more severe protein restriction that both sexes were affected (Woods et al. 2004). In the human, it has been shown that placental levels of 11β-hydroxysteroid dehydrogenase, which inactivates maternal cortisol before it enters the fetal circulation, is higher in mothers carrying female fetuses (Murphy et al. 2007). This means in any pregnancy, male fetuses may be exposed to higher levels of glucocorticoids than female fetuses.

8.8.2
Lifestyle Factors: A Congenital Nephron Deficit Renders the Kidney Susceptible to Secondary Insults

Being born with a low nephron endowment may increase the risk for developing adult disease such as hypertension or adult-onset diabetes; however, lifestyle influences may be the difference between one person getting these diseases and another individual not. It has been proposed that a congenital nephron deficit acts as a first hit on the kidney, rendering it susceptible to subsequent secondary postnatal hits (Nenov et al. 2000). Indeed, since loss of functioning glomeruli and nephrons through disease ultimately leads to end-stage renal disease, it is likely that disease progression will be accelerated when nephron endowment is already reduced at disease onset.

Reductions in nephron endowment are associated with compensatory hypertrophy of existing glomeruli, which increases glomerular filtration surface area (Black et al. 2004; Zimanyi et al. 2004). It is suggested, however, that as time goes by, hyperfiltration in hypertrophied glomeruli can ultimately lead to a breakdown in glomerular function and subsequent glomerular loss, thus perpetuating a vicious cycle (Hostetter 2003). In this regard, we frequently observe in our experimental studies an inverse correlation between glomerular number and glomerular size (Black et al. 2004; Zimanyi et al. 2004). Importantly, pronounced glomerular hypertrophy has been reported in the disease-prone kidneys of Australian Aborigines compared to non-Aboriginals (Young et al. 2000; Hoy et al. 2006); a reduced nephron endowment at birth and subsequent glomerulomegaly is linked to the high incidence of end-stage renal disease in this population. Based on the hypothesis proposed by Nenov et al. (2000), we postulate that a congenital nephron deficit acts as a first hit on the kidney. Initially, the kidney is able to adequately compensate through glomerular hypertrophy but with subsequent lifestyle insults (secondary hits) the kidney reaches its limits of compensation whereby subsequent decline in function ensues (Nenov et al. 2000). In support of this hypothesis, we and others have recently shown that the kidneys of rats with a congenital nephron deficit are more vulnerable to a secondary postnatal insult (Zimanyi et al. 2006; Plank et al. 2006). We have shown in IUGR rat offspring with a congenital nephron deficit that infusion of advanced glycation end-products (AGEs) from 20 to 24 weeks of age (at a concentration similar to that seen in the blood after 6 weeks of diabetes) leads to significant upregulation of TGFβ-1 and pro-collagen III in the kidney when compared to non-growth-restricted offspring. In addition, there is significant accumulation of AGEs within the kidneys (particularly in the glomeruli) in the rats with a congenital nephron deficit following AGE infusion compared with control kidneys. Since AGE formation is accelerated in the presence of elevated levels of circulating blood glucose, our findings suggest that a developmental nephron deficit may render the kidneys particularly susceptible to elevations in blood glucose levels. Importantly, we have also recently demonstrated that there is increased glycation of collagen in kidneys of the adult IUGR rat, leading to an increase in the formation of tissue AGEs (Zimanyi et al. 2006). The reason for the elevated levels of AGEs within the tissues of our adult IUGR rats is at this stage unknown. In addition, the kidney is the major organ responsible for AGE removal (Gugliucci and Bendayan 1996) and therefore a decline in renal function or a congenital reduction in nephron endowment (as occurs in IUGR) may facilitate chronic increases in circulating AGE levels. Alternatively, enhanced reactive oxygen species generation, leading to tissue oxidative stress, has been linked to tissue AGE formation independent of increased blood glucose levels (Bohlender et al. 2005).

In a rat model of maternal low protein exposure during pregnancy, exposure to a high-salt diet in adulthood was found to exacerbate the hypertension (Wood et al. 2004). In a recent study from our own laboratory, the male offspring of rats fed a low-protein diet throughout pregnancy, and who were maintained on this low-protein diet after birth, did not develop hypertension but did show a large increase in glomerular filtration rate when placed on a high-salt diet (Hoppe et al. 2007b). Interestingly, when male offspring of rats fed a normal (18%–20%) protein diet during pregnancy were cross-fostered during lactation onto dams fed a low-protein (8%) diet, there were beneficial effects on the kidneys at 1 year of age (Tarry-Adkins et al. 2007). These effects included maintenance of renal cortical telomere length and increased levels of antioxidant enzymes in the kidneys. Thus, protein restriction in lactation confers nephroprotective effects in the male rat and is associated with increased antioxidant expression. Hence, we propose that the onset and rate of progression of renal disease following a secondary renal insult in adulthood is accelerated in individuals who are born with a primary congenital nephron deficit when compared to those individuals whose kidneys are well endowed with nephrons at birth.

8.8.3
Models and Methodologies: Problems and Pitfalls

One of the difficulties in interpreting data from the DOHaD field is the wide variety of models utilized. The timing of any perturbation may have a huge impact on some organs but not others. The nature and duration of insult, along with species used, all have the potential to modify the disease outcome. This is particularly true in the maternal exposure to low-protein models; whilst the protein deficit in any two low-protein diets may be similar, often the remaining calories are provided from a variety of differing carbohydrate sources. These issues have been considered in detail in the excellent review article by Armitage et al. (2004). In a classic experiment, two different low-protein diets fed to pregnant rats were used within the one laboratory to remove differences in rat strains, laboratory conditions and blood pressure recording methodologies. It was found that only one of the low-protein maternal diets resulted in hypertension of the offspring, even though both diets produced reductions in neonatal birthweight (Langley-Evans 2000).

Another problem is the differing methodologies used. We have already stressed the importance of determining nephron endowment by unbiased stereological methods. Measurement of blood pressure is also a source of potential differences. Tail-cuff methods have long been recognized as being stressful and measurements obtained in this fashion are unlikely to represent true basal blood pressure. However, for studies where large numbers of animals require study at a variety of ages, tail cuff methods have some usefulness. In a model of prenatal malnutrition, blood pressure differences evident by tail cuff were absent when radiotelemetry was used (Tonkiss et al. 1998), suggesting an enhanced pressor response to stress but no differences in resting arterial pressure. Long-term measurements (over at least 24 h) using indwelling arterial catheters or telemetry probes are the optimal methods for measurement of basal blood pressure.

8.9
Unilateral Nephrectomy as a Model of Low Nephron Endowment

An argument that is often raised against a role for nephron number being of great importance in blood pressure regulation is the fact that healthy adults can donate a kidney for transplantation and not develop elevated blood pressure. The DOHaD hypothesis, however, suggests it is a low nephron endowment *from birth* that is important.

One of the difficulties in determining the importance of a low nephron number, and the associated renal changes in the development of adult disease, is that most in utero perturbations have effects on a variety of developing organs and systems. We have taken a unique approach to overcome this problem by establishing an ovine model of reduced nephron endowment. As discussed in Sect. 7.4, the sheep begins formation of the permanent (metanephric) kidney around day 27 of gestation and completes nephrogenesis around 130 days, some 3 weeks prior to birth, making the ovine fetus almost identical to the human fetus in terms of the timing of nephrogenesis. We have established a model of reduced nephron number by performing a unilateral nephrectomy (Uni-X) in sheep fetuses at 100 days of gestation (term = 150 days). This reduces overall nephron endowment without perturbation of the mother or placenta. Once delivered, Uni-X lambs were of normal birthweight and grew normally. Moderate elevations in blood pressure were demonstrated in the Uni-X female offspring at 6 months and at 1 and 2 years of age (Moritz et al. 2002b, 2005b); at 2 years of age, the increase in blood pressure was due to an increase in cardiac output; however, echocardiography showed no evidence of left ventricular hypertrophy at this age (Moritz et al. 2005b). The female Uni-X sheep had a reduced GFR and at 2 years of age showed a significantly attenuated GFR response to an amino acid infusion, indicating a reduced ability to excrete a protein load. At 2 years of age (but at 6 or 12 months) they also had increased plasma creatinine concentrations. More recently, we have shown that unilateral nephrectomy also results in hypertension and impaired cardiac function in male offspring (K.M. Moritz, unpublished data). Further studies in this model will contribute to our knowledge of how a low nephron number from birth results in cardiovascular disease.

8.10
Potential Underlying Mechanisms of Developmental Programming

8.10.1
Epigenetic Regulation

Through the phenomenon of epigenetics, patterns of gene expression can be changed independently of the existing DNA sequence. These modifications include cytosine methylation in the promoter region of genes, which silences transcription, and recruitment of chromatin remodelling (methylation, acetylation, phosphorylation of the histone proteins packaging DNA) complexes (Vickaryous and Whitelaw 2006; Waterland 2006; Ozanne and Constanci 2007). Imprinted genes are particularly vulnerable, as only the maternal or paternal copy of the gene is normally expressed (Fowden et al. 2006). There are now examples of epigenetic changes brought about in the embryo, fetus and neonate by maternal diet (Lillycrop et al. 2005; Drake et al. 2005; Kwong et al. 2006; Bogdarina et al. 2007), drugs such as cocaine (Zhang et al. 2007), endocrine disruptors (Chang et al. 2006) and, in the rat, maternal behaviour towards the neonate (Weaver et al. 2004, 2005, 2007; Meaney et al. 2007). So far, genes shown to be altered have been expressed in liver, brain, and adrenal but not the kidney. Nevertheless, as detailed in Sect. 3, numerous examples of altered gene expression have been found in the kidney as a consequence of some environmental event, particularly in the components of the RAS. It is most likely that these represent epigenetic changes brought about by the programming event, be it alterations in the protein content of the diet or exposure to excess natural or synthetic glucocorticoids (Moritz et al. 2005a).

8.10.2
Mitochondrial Dysfunction

Another potential molecular mechanism underlying the programming of adult diseases is mitochondrial dysfunction (McConnell 2006; Dickinson and Wintour 2007). There is evidence showing that mitochondrial DNA (mtDNA) mutations are linked to and precede the development of diabetes and hypertension in humans, implying a causal link (Song et al. 2001; Ballinger et al. 2002). Experimental manipulations of mtDNA in mice either in vitro in pancreatic cell lines or in vivo using conditional tissue-specific gene ablation approaches, show a direct link with altered mtDNA and altered pancreatic and cardiac function. Defective mtDNA-specific DNA polymerase has also been shown to lead to more rapid rates of ageing in mice (Trifunovic et al. 2004). In a rat model involving prenatal and suckling exposure to a diet rich in animal fat, which leads to whole body insulin resistance and pancreatic β-cell dysfunction in adulthood, there is evidence of reduced tissue mtDNA content and altered mitochondrial gene expression that precedes the insulin resistance (Taylor et al. 2005).

9
Future Directions

There are many areas ripe for research in the area of kidney development. With the current advances in technology available to image and examine the developing kidney along with transgenic animal technology, many new strategies can be employed to answer old questions as well as devising new ones. In the following sections, we consider a few very recent areas of interest in the field of renal development.

9.1
Clock Genes

It is now known that most cells in different organs have intrinsic clocks that regulate the functioning of the organ/system in a time-dependent fashion, mostly around a 24-h, circadian time frame. The level of expression of genes involved in these processes may change up to 20fold throughout the 24-h period (Boden and Kennaway 2006). These intrinsic clocks depend on the constant expression of a gene–*CLOCK* (circadian locomotor output cycle kaput)–whose protein product has to heterodimerize with that of a second gene, *BMAL1* (brain and muscle ARNT-like protein 1), to be effective. The dimer affects the expression of other timing genes such as the three *Period* (*Per*) genes and two *Cryptochrome* (*Cry*) genes. The protein products of the *Per* and *Cry* genes both have negative feedback effects on the primary genes and regulate the expression of other genes responsible for many metabolic, hormonal and other physiological functions (Kalsbeek et al. 2006; Boden and Kennaway 2006). These genes are autonomous, but they can be synchronized by a master clock in the suprachiasmatic nucleus (SCN) of the hypothalamus or by metabolic/other signals. Various disorders, particularly of early ageing and age-related pathologies have been seen in mice deficient in *BMAL1* (Kondratov et al. 2006), in Alzheimer's disease in humans (Wu et al. 2006) and in chromosome abnormalities in humans (De Leersnyder et al. 2006). In mice with the double knock-out of Cry1 and Cry2, mean arterial pressure was equal in both day and night periods, instead of being higher at night, and was higher at both time periods in the knock-out mice than the wild-type mice (Masuki et al. 2005). In addition, there were significant changes in baroreceptor sensitivity and response to adrenergic drugs. This establishes that abnormalities of some clock genes can produce hypertension.

The major zeitgeber or synchronizer receives input by light, for example, and then establishes a day/night cycle of behaviour or hormone release. One of the major systems regulated by the SCN is the circadian synthesis and release of the hormone melatonin by the pineal gland (Jilg et al. 2005). Melatonin concentrations rise with the onset of darkness and maternal melatonin can cross the placenta and entrain at least some of the circadian systems in the primate fetus (Torres-Farfan et al. 2006). Thus the fetus can tell the time even before it is born.

In the rat fetus, the circadian network is not fully developed before birth (Sumova et al. 2006) and exciting recent results suggest that some of the long-term consequences of alcohol abuse during pregnancy may result because the alcohol permanently altered the expression of some clock-gene mechanisms governing the expression of stress-related hormones in the brains of the adult offspring (Chen et al. 2006a). Adult male rats, whose dams had been given a diet in which 33% of calories came from ethanol during days 10–21 of pregnancy, demonstrate abnormal circadian expression of the mRNA for *proopiomelanocortin*, encoding the β-endorphin protein in the hypothalamus. These male offspring brains also had abnormalities in the *Per* genes of the SCN and the β-endorphin-containing neurons. This finding has implications beyond those of fetal alcohol exposure. With particular relevance to the kidney, it has been shown that the level of expression of the sodium transporter, NHE3, shows diurnal variation under the influence of clock genes (Saifur Rohman et al. 2005). Thus, it is of great importance in all developmental studies of renal gene expression that all tissues collected are obtained at the same time of day. Investigation of the expression of clock genes in the kidney may also provide some answers as to mechanisms of programming.

9.2
Renal Stem Cells and Renal Regeneration

Regenerative therapies for the treatment of chronic kidney disease has emerged in recent years as an exciting new research direction. It is well known that the kidney can undergo significant repair and regeneration following certain types of injury or disease (see Cochrane et al. 2005; Little et al. 2006). This repair may be the result of activity of resident kidney cells and/or non-renal cells, for example cells recruited from the bone marrow (reviewed by Anglani et al. 2004; Ricardo et al. 2005; Little et al. 2006). The adult kidney contains approximately 26 differentiated cell types. However, as described in Sect. 2.3, these cells arise from only two embryonic tissues: (1) the WD from which the UB and collecting duct system is derived; and (2) the MM which gives rise to the epithelial cells of the nephron, stroma, endothelial cells and smooth muscle cells. These two tissues (UB and MM) are believed to contain progenitor cells for these various cell types (reviewed in al-Awqati et al. 2002). However, the question remains whether these progenitors persist in the adult kidney and whether they play a role in the repair or regeneration of cell types following renal injury. Although adult renal stem cells located within the renal papilla, proximal tubules, and side population have been reported to express CD133, no definitive data exists that these cells contain all the characteristics of a stem cell such as pluripotency, self-renewal and clonogenicity (reviewed in Little et al. 2006).

9.3
New Renal Factors

It has recently been shown that the kidney produces a hormone called renalase that metabolizes catecholamines (Xu et al. 2005). In humans, renalase gene expression is highest in the kidney but is also detectable in the heart, skeletal muscle and small intestine. Plasma renalase concentrations are reduced in patients with end-stage renal disease. Administration of renalase lowers blood pressure and heart rate by metabolizing circulating catecholamines. Recent studies have shown abnormalities in the renalase pathway in animal models of chronic kidney disease and hypertension (Xu and Desir 2007). Therefore, in the future renalase may become an important therapeutic agent in the treatment of renal and cardiovascular disease.

In the last few years, a new component of the RAS, called ACE2, has been identified, and although tissue expression of ACE2 is now thought to be widespread, the kidney is a major site of production (reviewed in Hamming et al. 2007). ACE2 may play a pivotal role in controlling the balance between the vasoconstrictor effects of Ang II and the vasodilatory properties of other RAS components such as the angiotensin (1–7) peptide (Lazartigues et al. 2007). ACE2 has been implicated in cardiovascular and renal disease, diabetes, pregnancy, lung disease and, surprisingly, ACE2 also acts as a receptor for the SARS virus (Hamming et al. 2007; Lazartigues et al. 2007).

References

Abercrombie M (1946) Estimation of nuclear populations from microtomic sections. Anat Rec 94:239–247

Airik R, Bussen M, Singh MK, Petry M, Kispert A (2006) Tbx18 regulates the development of the ureteral mesenchyme. J Clin Invest 116:663–674

al-Awqati Q, Goldberg MR (1998) Architectural patterns in branching morphogenesis in the kidney. Kidney Int 54:1832–1842

Alcorn D, Maric C, McCausland J (1999) Development of the renal interstitium. Pediatr Nephrol 13:347–354

Alebiosu CO (2003) An update on 'progression promoters' in renal diseases. J Natl Med Assoc 95:30–42

Al-Bhalal L, Akhtar M (2005) Molecular basis of autosomal dominant polycystic kidney disease. Adv Anat Pathol 12:126–133

Amsalem H, Valsky DV, Yagel S, Celnikier DH, Anteby EY (2003) Effect of indomethacin on amniotic fluid prostaglandin and aldosterone levels in a fetus with Bartter syndrome. Prenat Diagn 23:431–433

Andreoli SP (2004) Acute renal failure in the newborn. Semin Perinatol 28:112–123

Armitage JA, Khan IY, Taylor PD, Nathanielsz PW, Poston L (2004) Developmental programming of the metabolic syndrome by maternal nutritional imbalance: how strong is the evidence from experimental models in mammals? J Physiol 561:355–377

Armitage JA, Taylor PD, Poston L (2005) Experimental models of developmental programming: Consequences of exposure to an energy rich diet during development. J Physiol 565:3–8

Awad H, el-Safty I, el-Barbary M, Imam S (2002) Evaluation of renal glomerular and tubular functional and structural integrity in neonates. Am J Med Sci 324:261–266

Bagby SP (2006) Developmental origins of hypertension: biology meets statistics. J Am Soc Nephrol 17:2356–2358

Ballinger SW, Patterson C, Knight-Lozano CA, Burow DL, Conklin CA, Hu Z, Reuf J, Horaist C, Lebovitz R, Hunter GC, McIntyre K, Runge MS (2002) Mitochondrial integrity and function in atherogenesis. Circulation 106:544–549

Barasch J, Yang J, Ware CB, Taga T, Yoshida K, Erdjument-Bromage H, Tempst P, Parravicini E, Malach S, Aranoff T, Oliver JA (1999) Mesenchymal to epithelial conversion in rat metanephros is induced by LIF. Cell 99:377–386

Barker DJ (2007) The origins of the developmental origins theory. J Intern Med 261:412–417

Barker DJ, Bagby SP (2005) Developmental antecedents of cardiovascular disease: a historical perspective. J Am Soc Nephrol 16:2537–2544

Barker DJ, Osmond C (1986) Infant mortality, childhood nutrition, and ischemic heart disease in England and Wales. Lancet 1:1077–1081

Barker DJP (1998) Mothers, babies and health in later life,2nd edn. Churchill Livingstone, Edinburgh

Barker DJP, Eriksson JG, Forsen T, Osmond C (2002) Fetal origins of adult disease: strength of effects and biological basis. Int J Epidemiol 31:1235–1239

Basson MA, Akbulut S, Watson-Johnson J, Simon R, Carroll TJ, Shakya R, Gross I, Martin GR, Lufkin T, McMahon AP, Wilson PD, Costantini FD, Mason IJ, Licht JD (2005) Sprouty1 is a critical regulator of GDNF/RET-mediated kidney induction. Dev Cell 8:229–239

Batourina E, Gim S, Bello N, Shy M, Clagett-Dame M, Srinivas S, Costantini F, Mendelsohn C (2001) Vitamin A controls epithelial/mesenchymal interactions through Ret expression. Nat Genet 27:74–78

Bertram C, Trowern AR, Copin N, Jackson AA, Whorwood CB (2001) The maternal diet during pregnancy programs altered expression of the glucocorticoid receptor and type 2 11beta-hydroxysteroid dehydrogenase: potential molecular mechanisms underlying the programming of hypertension in utero. Endocrinology 142:2841–2853

Bertram JF (1995) Analyzing renal glomeruli with the new stereology. Int Rev Cytol 161:111–172

Bertram JF (2001) Counting in the kidney. Kidney Int 59:792–796

Bertram JF, Soosaipillai MC, Ricardo SD, Ryan GB (1992) Total numbers of glomeruli and individual glomerular cell types in the normal rat kidney. Cell Tissue Res 270:37–45

Beutler KT, Masilamani S, Turban S, Nielsen J, Brooks HL, Ageloff S, Fenton RA, Packer RK, Knepper MA (2003) Long-term regulation of ENaC expression in kidney by angiotensin II. Hypertension 41:1143–1150

Black MJ, Briscoe TA, Constantinou M, Kett MM, Bertram JF (2004) Is there an association between level of adult blood pressure and nephron number or renal filtration surface area? Kidney Int 65:582–588

Black MJ, Wang Y, Bertram JF (2002) Nephron endowment and renal filtration surface area in young spontaneously hypertensive rats. Kidney Blood Press Res 25:20–26

Bloomfield FH, Oliver MH, Giannoulias CD, Gluckman PD, Harding JE, Challis JR (2003) Brief undernutrition in late-gestation sheep programs the hypothalamic-pituitary-adrenal axis in adult offspring. Endocrinology 144:2933–2940

Boden MJ, Kennaway DJ (2006) Circadian rhythms and reproduction. Reproduction 132:379–392

Bogdarina I, Welham S, King PJ, Burns SP, Clark AJ (2007) Epigenetic modification of the renin-angiotensin system in the fetal programming of hypertension. Circ Res 100:520–526

Bohlender J, Franke S, Sommer M, Stein G (2005) Advanced glycation end products: a possible link to angiotensin in an animal model. Ann N Y Acad Sci 1043:681–684

Bouchard M (2004) Transcriptional control of kidney development. Differentiation 72:295–306

Bouchard M, Souabni A, Mandler M, Neubuser A, Busslinger M (2002) Nephric lineage specification by Pax2 and Pax8. Genes Dev 16:2958–2970

Boyle S, de Caestecker M (2006) Role of transcriptional networks in coordinating early events during kidney development. Am J Physiol Renal Physiol 291:F1–F8

Brenner BM, Anderson S (1992) The inter-relationships among filtration surface area, blood pressure and chronic renal disease. J Cardiovasc Pharmacol 19:S1–S6

Brenner BM, Garcia DL, Anderson S (1988) Glomeruli and blood pressure: less of one, more of the other? Am J Hypertens 1:335–347

Bullock SL, Fletcher JM, Beddington RS, Wilson VA (1998) Renal agenesis in mice homozygous for a gene trap mutation in the gene encoding heparan sulfate 2-sulfotransferase. Genes Dev 12:1894–1906

Burrow CR (2000) Regulatory molecules in kidney development. Pediatr Nephrol 14:240–253

Butkus A, Albiston A, Alcorn D, Giles M, McCausland J, Moritz K, Zhuo J, Wintour EM (1997) Ontogeny of angiotensin II receptors, types 1 and 2, in ovine mesonephros and metanephros. Kidney Int 52:628–636

Cacalano G, Farinas I, Wang LC, Hagler K, Forgie A, Moore M, Armanini M, Phillips H, Ryan AM, Reichardt LF, Hynes M, Davies A, Rosenthal A (1998) GFRalpha1 is an essential receptor component for GDNF in the developing nervous system and kidney. Neuron 21:53–62

Cain JE, Bertram JF (2006) Ureteric branching morphogenesis in BMP4 heterozygous mutant mice. J Anat 209:745–755

Cain JE, Nion T, Jeulin D, Bertram JF (2005) Exogenous BMP-4 amplifies asymmetric ureteric branching in the developing mouse kidney in vitro. Kidney Int 67:420–431

Carev D, Krnic D, Saraga M, Sapunar D, Saraga-Babic M (2006) Role of mitotic, pro-apoptotic and anti-apoptotic factors in human kidney development. Pediatr Nephrol 21:627–636

Carroll TJ, Park JS, Hayashi S, Majumdar A, McMahon AP (2005) Wnt9b plays a central role in the regulation of mesenchymal to epithelial transitions underlying organogenesis of the mammalian urogenital system. Dev Cell 9:283–292

Caruana G, Cullen-McEwen L, Nelson AL, Kostoulias X, Woods K, Gardiner B, Davis MJ, Taylor DF, Teasdale RD, Grimmond SM, Little MH, Bertram JF (2006a) Spatial gene expression in the T-stage mouse metanephros. Gene Expr Patterns 6:807–825

Caruana G, Young RJ, Bertram JF (2006b) Imaging the embryonic kidney. Nephron Exp Nephrol 103:e62–e68

Cataldi L, Leone R, Moretti U, De Mitri B, Fanos V, Ruggeri L, Sabatino G, Torcasio F, Zanardo V, Attardo G, Riccobene F, Martano C, Benini D, Cuzzolin L (2005) Potential risk factors for the development of acute renal failure in preterm newborn infants: a case-control study. Arch Dis Child Fetal Neonatal Ed 90:F514–F519

Challen G, Gardiner B, Caruana G, Kostoulias X, Martinez G, Crowe M, Taylor DF, Bertram J, Little M, Grimmond SM (2005) Temporal and spatial transcriptional programs in murine kidney development. Physiol Genomics 23:159–171

Challen GA, Bertoncello I, Deane JA, Ricardo SD, Little MH (2006) Kidney side population reveals multilineage potential and renal functional capacity but also cellular heterogeneity. J Am Soc Nephrol 17:1896–1912

Challen GA, Martinez G, Davis MJ, Taylor DF, Crowe M, Teasdale RD, Grimmond SM, Little MH (2004) Identifying the molecular phenotype of renal progenitor cells. J Am Soc Nephrol 15:2344–2357

Chan T, Asashima M (2006) Growing kidney in the frog. Nephron Exp Nephrol 103:e81–e85

Chang HS, Anway MD, Rekow SS, Skinner MK (2006) Transgenerational epigenetic imprinting of the male germline by endocrine disruptor exposure during gonadal sex determination. Endocrinology 147:5524–5541

Chen CP, Kuhn P, Advis JP, Sarkar DK (2006a) Prenatal ethanol exposure alters the expression of period genes governing the circadian function of β-endorphin neurons in the hypothalamus. J Neurochem 97:1026–1033

Chen H, Lun Y, Ovchinnikov D, Kokubo H, Oberg KC, Pepicelli CV, Gan L, Lee B, Johnson RL (1998) Limb and kidney defects in Lmx1b mutant mice suggest an involvement of LMX1B in human nail patella syndrome. Nat Genet 19:51–55

Chen L, Al-Awqati Q (2005) Segmental expression of Notch and Hairy genes in nephrogenesis. Am J Physiol Renal Physiol 288:F939–F952

Chen N, Aleksa K, Woodland C, Rieder M, Koren G (2006b) Ontogeny of drug elimination by the human kidney. Pediatr Nephrol 21:160–168

Chen Y, Lasaitiene D, Gabrielsson BG, Carlsson LM, Billig H, Carlsson B, Marcussen N, Sun XF, Friberg P (2004) Neonatal losartan treatment suppresses renal expression of molecules involved in cell-cell and cell-matrix interactions. J Am Soc Nephrol 15:1232–1243

Chevalier RL (2004) Perinatal obstructive nephropathy. Semin Perinatol 28:124–131

Cho EA, Patterson LT, Brookhiser WT, Mah S, Kintner C, Dressler GR (1998) Differential expression and function of cadherin-6 during renal epithelium development. Development 125:803–812

Christian P (2003) Micronutrients and reproductive health issues: an international perspective. J Nutr 133:1969S–1973S

Clark AT, Bertram JF (1999) Molecular regulation of nephron endowment. Am J Physiol Renal Physiol 276:485–497

Clark AT, Young RJ, Bertram JF (2001) In vitro studies on the roles of transforming growth factor-beta 1 in rat metanephric development. Kidney Int 59:1641–1653

Cockey CD (2004) Prematurity hits record high. AWHONN Lifelines 8:104–105

Costantini F, Shakya R (2006) GDNF/Ret signaling and the development of the kidney. Bioessays 28:117–127

Crowe C, Dandekar P, Fox M, Dhingra K, Bennet L, Hanson MA (1995) The effects of anaemia on heart, placenta and body weight, and blood pressure in fetal and neonatal rats. J Physiol 488:515–519

Cullen-McEwen LA, Drago J, Bertram JF (2001) Nephron endowment in glial cell line-derived neurotrophic factor (GDNF) heterozygous mice. Kidney Int 60:31–36

Cullen-McEwen LA, Caruana G, Bertram JF (2005) The where, what and why of the developing renal stroma. Nephron Exp Nephrol 99:e1–e8

Cullen-McEwen LA, Fricout G, Harper IS, Jeulin D, Bertram JF (2002) Quantitation of 3D ureteric branching morphogenesis in cultured embryonic mouse kidney. Int J Dev Biol 46:1049–1055

Curhan GC, Willett WC, Rimm EB, Spiegelman D, Ascherio AL, Stampfer MJ (1996) Birth weight and adult hypertension, diabetes mellitus, and obesity in US men. Circulation 94:3246–3250

Dagan A, Gattineni J, Cook V, Baum M (2007) Prenatal programming of rat proximal tubule Na^+/H^+ exchanger by dexamethasone. Am J Physiol Regul Integr Comp Physiol 292: R1230–R1235

Dahl U, Sjodin A, Larue L, Radice GL, Cajander S, Takeichi M, Kemler R, Semb H (2002) Genetic dissection of cadherin function during nephrogenesis. Mol Cell Biol 22: 1474–1487

Dakovic-Bjelakovic M, Vlajkovic S, Cukuranovic R, Antic S, Bjelakovic G, Mitic D (2005) Quantitative analysis of the nephron during human fetal kidney development. Vojnosanit Pregl 62:281–286

Davies J (2001) Intracellular and extracellular regulation of ureteric bud morphogenesis. J Anat 198:257–264

Davies JA, Ladomery M, Hohenstein P, Michael L, Shafe A, Spraggon L, Hastie N (2004) Development of an siRNA-based method for repressing specific genes in renal organ culture and its use to show that the Wt1 tumour suppressor is required for nephron differentiation. Hum Mol Genet 13:235–246

De Leersnyder H, Claustrat B, Munnich A, Verloes A (2006) Circadian rhythm disorder in a rare disease-Smith-Magenis syndrome. Mol Cell Endocrinol 252:88–91

de Onis M, Blossner M, Villar J (1998) Levels and patterns of intrauterine growth retardation in developing countries. Eur J Clin Nutr 52:S5–S15

DeHoff RT, Rhines FN (1961) Determination of number of particles per unit, from measurements made on random plane sections: The general cylinder and the ellipsoid. Trans Metall Soc AIME 221:975–982

Denton K, Baylis C (2007) Physiological and molecular mechanisms governing sexual dimorphism of kidney, cardiac, and vascular function. Am J Physiol Regul Integr Comp Physiol 292:R697–R699

Dickinson H, Wintour EM (2007) Can life before birth affect health ever after? Current Women's Health Review 3:79–88

Dickinson H, Walker DW, Cullen-McEwen L, Wintour EM, Moritz K (2005) The spiny mouse (*Acomys cahirinus*) completes nephrogenesis before birth. Am J Physiol Renal Physiol 289:F273–279. Erratum in: Am J Physiol Renal Physiol (2005) 289:F1386

Dickinson H, Walker DW, Wintour EM, Moritz K (2007) Maternal dexamethasone treatment at midgestation reduces nephron number and alters renal gene expression in the fetal spiny mouse. Am J Physiol Regul Integr Comp Physiol 292:R453–R461

Dodic M, Hantzis V, Duncan J, Rees S, Koukoulas I, Johnson K, Wintour EM, Moritz K (2002) Programming effects of short prenatal exposure to cortisol. FASEB J 16:1017–1026

Dodic M, May CN, Wintour EM, Coghlan JP (1998) An early prenatal exposure to excess glucocorticoid leads to hypertensive offspring in sheep. Clin Sci 94:149–155

Donnelly S (2001) Why is erythropoietin made in the kidney? The kidney functions as a critmeter. Am J Kidney Dis 38:415–425

Douglas-Denton R, Moritz KM, Bertram JF, Wintour EM (2002) Compensatory renal growth after unilateral nephrectomy in the ovine fetus. J Am Soc Nephrol 13:406–410

Douglas-Denton RN, McNamara BJ, Hoy WE, Hughson MD, Bertram JF (2006) Does nephron number matter in the development of kidney disease? Ethn Dis 16:S2–40–45

Drake AJ, Walker BR, Seckl JR (2005) Intergenerational consequences of fetal programming by in utero exposure to glucocorticoids in rats. Am J Physiol Regul Integr Comp Physiol 288:R34–R38

Dressler GR, Deutsch U, Chowdhury K, Nornes HO, Gruss P (1990) Pax2, a new murine paired-box-containing gene and its expression in the developing excretory system. Development 109:787–795

Drukker A, Guignard J (2002) Renal aspects of the term and preterm infant: a selective update. Curr Opin Pediatr 14:175–182

Drummond IA, Majumdar A (2003) The pronephric glomus and vasculature. In: Vize PD, Woolf AS, Bard JBL (eds) The kidney: from normal development to congenital disease. Academic Press, London, pp 61–73

Dudley AT, Lyons KM, Robertson EJ (1995) A requirement for bone morphogenetic protein-7 during development of the mammalian kidney and eye. Genes Dev 9:2795–2807

Dunn NR, Winnier GE, Hargett LK, Schrick JJ, Fogo AB, Hogan BL (1997) Haploinsufficient phenotypes in Bmp4 heterozygous null mice and modification by mutations in Gli3 and Alx4. Dev Biol 188:235–247

Durbec P, Marcos-Gutierrez CV, Kilkenny C, Grigoriou M, Wartiowaara K, Suvanto P, Smith D, Ponder B, Costantini F, Saarma M, Sariola H, Pachnis V (1996) GDNF signalling through the Ret receptor tyrosine kinase. Nature 381:789–793

Ekblom P (1992) Renal development. In: Seldin DW, Giebisch G (eds) Kidney physiology and pathophysiology. Raven Press, New York, pp 475–501

Ekblom P, Weller A (1991) Ontogeny of tubulointerstitial cells. Kidney Int 39:394–400

Enomoto H, Araki T, Jackman A, Heuckeroth RO, Snider WD, Johnson EM Jr, Milbrandt J (1998) GFR alpha1-deficient mice have deficits in the enteric nervous system and kidneys. Neuron 21:317–324

Eremina V, Wong MA, Cui S, Schwartz L, Quaggin SE (2002) Glomerular-specific gene excision in vivo. J Am Soc Nephrol 13:788–793

Eriksson JG, Forsén T, Tuomilehto J, Winter PD, Osmond C, Barker DJ (1999) Catch-up growth in childhood and death from coronary heart disease: longitudinal study. BMJ 318:427–431

Eskild-Jensen A, Frokiaer J, Djurhuus JC, Jorgensen TM, Nyengaard JR (2002) Reduced number of glomeruli in kidneys with neonatally induced partial ureteropelvic obstruction in pigs. J Urol 167:1435–1439

Esquela AF, Lee SJ (2003) Regulation of metanephric kidney development by growth/differentiation factor 11. Dev Biol 257:356–370

Fall CHD (2001) Non-industrialised countries and affluence: relationship with type 2 diabetes. Br Med Bull 60:33–50

Fall CHD, Sachdev HS (2006) Developmental origins of health and disease: implications for developing countries. In: Gluckman PD, Hanson MA (eds) Developmental origins of health and disease. Cambridge University Press, Cambridge, pp 456–471

Floderus S (1944) Untersuchungen uber den bau der menschlichen hypophyse mit besonderer berucksichtigung der quantitativen mikromorphologischen verhaltnisse. Acta Pathol Microbiol Scand 53:1–26

Fowden AL, Sibley C, Reik W, Constancia M (2006) Imprinted genes, placental development and fetal growth. Horm Res 65:50–58

Fricout G, Cullen-McEwen LA, Harper IS, Jeulin D, Bertram JF (2001) A quantitative method for analyzing 3-D branching in embryonic kidneys: development of a technique and preliminary data. Image Anal Stereol 20:36–41

Fricout G, Jeulin D, Cullen-McEwen LA, Harper IS, Bertram JF (2002) 3D-skeletonization of ureteric trees in developing kidneys. In: Talbot H, Beare R (eds) Mathematical morphology: proceedings of the VIth international symposium ISSM 2002. CSIRO Publishing, Collingwood, pp 157–164

Gallini F, Maggio L, Romagnoli C, Marrocco G, Tortorolo G (2000) Progression of renal function in preterm neonates with gestational age < or = 32 weeks. Pediatr Nephrol 15:119–124

Gambling L, Dunford S, Wilson CA, McArdle HJ, Baines DL (2004) Estrogen and progesterone regulate alpha, beta, and gammaENaC subunit mRNA levels in female rat kidney. Kidney Int 65:1774–1781

Gilbert T, Merlet-Benichou C (2000) Retinoids and nephron mass control. Pediatr Nephrol 14:1137–1144

Gluckman PD, Hanson MA (2006) The developmental origins of health and disease: an overview. In: Gluckman PD, Hanson MA (eds) Developmental origins of health and disease. Cambridge University Press, Cambridge, pp 1–5

Godfrey K (2006) The 'developmental origins' hypothesis: epidemiology. In: Gluckman PD, Hanson MA (eds) Developmental origins of health and disease. Cambridge University Press, Cambridge, pp 6–32

Gopalakrishnan GS, Gardner DS, Dandrea J, Langley-Evans SC, Pearce S, Kurlak LO, Walker RM, Seetho IW, Keisler DH, Ramsay MM, Stephenson T, Symonds ME (2005) Influence of maternal pre-pregnancy body composition and diet during early-mid pregnancy on cardiovascular function and nephron number in juvenile sheep. Br J Nutr 94:938–947

Grieshammer U, Le Ma, Plump AS, Wang F, Tessier-Lavigne M, Martin GR (2004) SLIT2-mediated ROBO2 signalling restricts kidney induction to a single site. Dev Cell 6:709–717

Grobstein C (1956) Transfilter induction of tubules in mouse metanephrogenic mesenchyme. Exp Cell Res 10:424–440

Grote D, Souabni A, Busslinger M, Bouchard M (2006) Pax 2/8-regulated Gata 3 expression is necessary for morphogenesis and guidance of the nephric duct in the developing kidney. Development 133:53–61

Gubhaju L, Black MJ (2005) The baboon as a good model for studies of human kidney development. Pediatr Res 58:505–509

Gubhaju L, Zulli A (2005) The effect of pre-term birth on nephrogenesis. Pediatr Res 58:1065

Gugliucci A, Bendayan M (1996) Renal fate of circulating advanced glycated end products (AGE): evidence for reabsorption and catabolism of AGE-peptides by renal proximal tubular cells. Diabetologia 39:149–160

Guillery E (1997) Fetal and neonatal nephrology. Curr Opin Pediatr 9:148–153

Hadjantonakis AK, Dickinson ME, Fraser SE, Papaioannou VE (2003) Technicolour transgenics: imaging tools for functional genomics in the mouse. Nat Rev Genet 4:613–625

Hadjantonakis AK, Macmaster S, Nagy A (2002) Embryonic stem cells and mice expressing different GFP variants for multiple non-invasive reporter usage within a single animal. BMC Biotechnol 2:11

Hales CN, Ozanne SE (2003) For debate: fetal and early postnatal growth restriction lead to diabetes, the metabolic syndrome and renal failure. Diabetologia 46:1013–1019

Hamming I, Cooper ME, Haagmans BL, Hooper NM, Korstanje R, Osterhaus AD, Timens W, Turner AJ, Navis G, van Goor H (2007) The emerging role of ACE2 in physiology and disease. J Pathol 212:1–11

Harper IS, Cullen LA, Bertram JF (2001) Quantitative studies of branching morphogenesis in the developing kidney. In: Fleury V, Gouyet J-F, Leonetti M (eds) Branching in nature: dynamics and morphogenesis of branching structures, from cells to river networks. Springer-Verlag, Berlin and EDP Sciences, Les Ulis, pp 243–249

Harvey N, Cooper C (2004) The developmental origins of osteoporotic fracture. J Br Menopause Soc 10:14–15

Haskell MJ, Pandey P, Graham JM, Peerson JM, Shrestha RK, Brown KH (2005) Recovery from impaired dark adaptation in nightblind pregnant Nepali women who receive small daily doses of vitamin A as amaranth leaves, carrots, goat liver, vitamin A-fortified rice, or retinyl palmitate. Am J Clin Nutr 81:461–471

Hatini V, Huh SO, Herzlinger D, Soares VC, Lai E (1996) Essential role of stromal mesenchyme in kidney morphogenesis revealed by targeted disruption of winged helix transcription factor BF-2. Genes Dev 10:1467–1478

Hellmich HL, Kos L, Cho ES, Mahon KA, Zimmer A (1996) Embryonic expression of glial cell-line derived neurotrophic factor (GDNF) suggests multiple developmental roles in neural differentiation and epithelial-mesenchymal interactions. Mech Dev 54:95–105

Hensey C, Dolan V, Brady HR (2002) The Xenopus pronephros as a model system for the study of kidney development and pathophysiology. Nephrol Dial Transplant 17:73–74

Hinchliffe SA, Lynch MR, Sargent PH, Howard CV, Van Velzen D (1992) The effect of intrauterine growth retardation on the development of renal nephrons. Br J Obstet Gynaecol 99:296–301

Hinchliffe SA, Sargent PH, Howard CV, Chan YF, Van Velzen D (1991) Human intrauterine renal growth expressed in absolute number of glomeruli assessed by the dissector method and Cavalieri principle. Lab Invest 64:777–784

Hoekstra RE, Ferrara TB, Couser RJ, Payne NR, Connett JE (2004) Survival and long-term neurodevelopmental outcome of extremely premature infants born at 23–26 weeks' gestational age at a tertiary center. Pediatrics 113:e1–e6

Holtbäck U, Aperia AC (2003) Molecular determinants of sodium and water balance during early human development. Semin Neonatol 8:291–299

Hoppe CC, Evans RG, Bertram JF, Moritz KM (2007a) Effects of dietary protein restriction on nephron number in the mouse. Am J Physiol Regul Integr Comp Physiol 292:R1768–R1774

Hoppe CC, Evans RG, Moritz KM, Cullen-McEwen LA, Fitzgerald SM, Dowling J, Bertram JF (2007b) Combined prenatal and postnatal protein restriction influences adult kidney structure, function, and arterial pressure. Am J Physiol Regul Integr Comp Physiol 292: R462–R469

Hostetter TH (2003) Hyperfiltration and glomerulosclerosis. Semin Nephrol 23:194–199

Hoy WE, Douglas-Denton RN, Hughson MD, Cass A, Johnson K, Bertram JF (2003) A stereological study of glomerular number and volume: preliminary findings in a multiracial study of kidneys at autopsy. Kidney Int (83):S31–S37

Hoy WE, Hughson MD, Bertram JF, Douglas-Denton R, Amann K (2005) Nephron number, hypertension, renal disease, and renal failure. J Am Soc Nephrol 16:2557–2564

Hoy WE, Hughson MD, Singh GR, Douglas-Denton R, Bertram JF (2006) Reduced nephron number and glomerulomegaly in Australian Aborigines: a group at high risk for renal disease and hypertension. Kidney Int 70:104–110

Hoy WE, Rees M, Kile E, Mathews JD, Wang Z (1999) A new dimension to the Barker hypothesis: low birthweight and susceptibility to renal disease. Kidney Int 56:1072–1077

Hughson M, Farris AB 3rd, Douglas-Denton R, Hoy WE, Bertram JF (2003) Glomerular number and size in autopsy kidneys: the relationship to birth weight. Kidney Int 63:2113–2122

Hughson MD, Douglas-Denton R, Bertram JF, Hoy WE (2006) Hypertension, glomerular number, and birth weight in African Americans and white subjects in the southeastern United States. Kidney Int 69:671–678

Ichihara A, Kobori U, Nishiyama A, Navar LG (2004) Renal renin-angiotensin system. Contrib Nephrol 143:117–130

Iosipiv IV, Schroeder M (2003) A role for angiotensin II AT1 receptors in ureteric bud cell branching. Am J Physiol Renal Physiol 285:F199–F207

Itäranta P, Lin Y, Perasaari J, Roel G, Destree O, Vainio S (2002) Wnt-6 is expressed in the ureter bud and induces kidney tubule development in vitro. Genesis 32:259–268

Ito H, Wang J, Strandhoy JW, Rose JC (2001) Importance of the renal nerves for basal and stimulated renin mRNA levels in fetal and adult ovine kidneys. J Soc Gynecol Invest 8:327–333

Jain S, Suarez AA, McGuire J, Liapis H (2007) Expression profiles of congenital renal dysplasia reveal new insights into renal development and disease. Pediatr Nephrol 22:962–974

James RG, Kamei CN, Wang Q, Jiang R, Schultheiss TM (2006) Odd-skipped related 1 is required for development of the metanephric kidney and regulates formation and differentiation of kidney precursor cells. Development 133:2995–3004

Jilg A, Moek J, Weaver DR, Korf H-W, Stehle JH, von Gall C (2005) Rhythms in clock proteins in the mouse pars tuberalis depend on MT1 melatonin receptor signalling. Eur J Neurosci 22:2845–2854

Kalatzis V, Sahly I, El-Amraoui A, Petit C (1998) Eya1 expression in the developing ear and kidney: towards the understanding of the pathogenesis of Branchio-Oto-Renal (BOR) syndrome. Dev Dyn 213:486–499

Kalsbeek A, Perreau-Lenz S, Buijs RM (2006) A network of (autonomic) clock outputs. Chronobiol Inte 23:521–535

Kanwar YS, Ota K, Yang Q, Wada J, Kashihara N, Tian Y, Wallner EI (1999) Role of membrane-type matrix metalloproteinase 1 (MT-1-MMP), MMP-2, and its inhibitor in nephrogenesis. Am J Physiol 277:F934–F947

Karavanova ID, Dove LF, Resau JH, Perantoni AO (1996) Conditioned medium from a rat ureteric bud cell line in combination with bFGF induces complete differentiation of isolated metanephric mesenchyme. Development 122:4159–4167

Keijzer Veen MG, Schrevel M, Finken MJ, Dekker FW, Nauta J, Hille ET, Frolich M, van der Heijden BJ, Dutch POPS-19 Collaborative Study Group (2005) Microalbuminuria and lower glomerular filtration rate at young age in subjects born very premature and after intrauterine growth restriction. J Am Soc Nephrol 16:2762–2768

Keller G, Zimmer G, Mall G, Ritz E, Amann K (2003) Nephron number in patients with primary hypertension. N Engl J Med 348:101–108

Khorram O, Khorram N, Momeni M, Han G, Halem J, Desai M, Ross MG (2007) Maternal undernutrition inhibits angiogenesis in the offspring: a potential mechanism of programmed hypertension. Am J Physiol Regul Integr Comp Physiol 293:R745–R753

Kirshon B, Moise KJ, Wasserstrum N, Ou C, Huhta JC (1988) Influence of short-term indomethacin therapy on fetal urine output. Obstet Gynecol 72:51–53

Kitraki E, Kittas C, Stylianopoulou F (1997) Glucocorticoid receptor gene expression during rat embryogenesis. An in situ hybridization study. Differentiation 62:21–31

Kiyozumi D, Sugimoto N, Sekiguchi K (2006) Breakdown of the reciprocal stabilization of QBRICK/Frem1, Fras1, and Frem2 at the basement membrane provokes Fraser syndrome-like defects. Proc Natl Acad Sci U S A 103:11981–11986

Kloth S, Aigner J, Schmidbauer A, Minuth WW (1994) Interrelationship of renal vascular development and nephrogenesis. Cell Tissue Res 277:247–257

Knepper MA, Kim GH, Masilamani S (2003) Renal tubule sodium transporter abundance profiling in rat kidney: response to aldosterone and variations in NaCl intake. Ann N Y Acad Sci 986:562–569

Kobayashi H, Kawakami K, Asashima M, Nishinakamura R (2007) Six1 and Six4 are essential for Gdnf expression in the metanephric mesenchyme and ureteric bud formation, while Six1 deficiency alone causes mesonephric-tubule defects. Mech Dev 124:290–303

Kolatsi-Joannou M, Li XZ, Suda T, Yuan HT, Woolf AS (2001) Expression and potential role of angiopoietins and Tie-2 in early development of the mouse metanephros. Dev Dyn 222:120–126

Kondo S, Scheef EA, Sheibani N, Sorenson CM (2007) PECAM-1 isoform-specific regulation of kidney endothelial cell migration and capillary morphogenesis. Am J Physiol Cell Physiol 292:C2070–C2083

Kondratov RV, Kondratova AA, Gorbacheva VY, Vykhovanets OV, Antoch MP (2006) Early aging and age-related pathologies in mice deficient in BMAL1, the core component of the circadian clock. Genes Dev 20:1868–1873

Korgun ET, Dohr G, Desoye G, Demir R, Kayisli UA, Hahn T (2003) Expression of insulin, insulin-like growth factor I and glucocorticoid receptor in rat uterus and embryo during decidualization, implantation and organogenesis. Reproduction 125:75–84

Krege JH, John SW, Langenbach LL, Hodgin JB, Hagaman JR, Bachman ES, Jennette JC, O'Brien DA, Smithies O (1995) Male-female differences in fertility and blood pressure in ACE-deficient mice. Nature 375:146–148

Kreidberg JA (2003) Podocyte differentiation and glomerulogenesis. J Am Soc Nephrol 14:806–814

Kreidberg JA, Sariola H, Loring JM, Maeda M, Pelletier J, Housman D, Jaenisch R (1993) WT-1 is required for early kidney development. Cell 74:679–691

Kume T, Deng K, Hogan BL (2000) Murine forkhead/winged helix genes Foxc1 (Mf1) and Foxc2 (Mfh1) are required for the early organogenesis of the kidney and urinary tract. Development 127:1387–1395

Kwong WY, Miller DJ, Ursell E, Wild AE, Wilkins AP, Osmond C, Anthony FW, Fleming TP (2006) Imprinted gene expression in the rat embryo-fetal axis is altered in response to periconceptional maternal low protein diet. Reproduction 132:265–277

Lackland DT, Bendall HE, Osmond C, Egan BM, Barker DJ (2000) Low birth weights contribute to high rates of early-onset chronic renal failure in the Southeastern United States. Arch Intern Med 160:1472–1476

Lacoste M, Cai Y, Guicharnaud L, Mounier F, Dumez Y, Bouvier R, Dijoud F, Gonzales M, Chatten J, Delezoide AL, Daniel L, Joubert M, Laurent N, Aziza J, Sellami T, Amar HB, Jarnet C, Frances AM, Daikha-Dahmane F, Coulomb A, Neuhaus TJ, Foliguet B, Chenal P, Marcorelles P, Gasc JM, Corvol P, Gubler MC (2006) Renal tubular dysgenesis, a not uncommon autosomal recessive disorder leading to oligohydramnios: role of the renin-angiotensin system. J Am Soc Nephrol 17:2253–2263

Lane RH, Chandorkar AK, Flozak AS, Simmons RA (1998) Intrauterine growth retardation alters mitochondrial gene expression and function in fetal and juvenile rat skeletal muscle. Pediatr Res 43:563–570

Langley-Evans SC (2000) Critical differences between two low protein diet protocols in the programming of hypertension in the rat. Int J Food Sci Nutr 51:11–17

Langley-Evans SC, Welham SJ, Jackson AA (1999) Fetal exposure to a maternal low protein diet impairs nephrogenesis and promotes hypertension in the rat. Life Sci 64:965–974

Lasaitiene D, Chen Y, Adams MA, Friberg P (2006) Further insights into the role of angiotensin II in kidney development. Clin Physiol Funct Imaging 26:197–204

Lasaitiene D, Chen Y, Guron G, Marcussen N, Tarkowski A, Telemo E, Friberg P (2003) Perturbed medullary tubulogenesis in neonatal rat exposed to renin-angiotensin system inhibition. Nephrol Dial Transplant 18:2534–2541

Lazartigues E, Feng Y, Lavoie JL (2007) The two fACEs of the tissue renin-angiotensin systems: implication in cardiovascular diseases. Curr Pharm Des 13:1231–1245

Leeson TS (1959) An electron microscopic study of the mesonephros and metanephros of the rabbit. J Anat 105:165–195

Lelievre-Pegorier M, Vilar J, Ferrier ML, Moreau E, Freund N, Gilbert T, Merlet-Benichou C (1998) Mild vitamin A deficiency leads to inborn nephron deficit in the rat. Kidney Int 54:1455–1462

Levinson RS, Batourina E, Choi C, Vorontchikhina M, Kitajewski J, Mendelsohn CL (2005) Foxd1-dependent signals control cellularity in the renal capsule, a structure required for normal renal development. Development 132:529–539

Li Z, Stuart RO, Qiao J, Pavlova A, Bush KT, Pohl M, Sakurai H, Nigam SK (2000) A role for Timeless in epithelial morphogenesis during kidney development. Proc Natl Acad Sci U S A 97:10038–10043

Lillycrop KA, Phillips ES, Jackson AA, Hanson MA, Burdge GC (2005) Dietary protein restriction of pregnant rats induces and folic acid supplementation prevents epigenetic modification of hepatic gene expression in the offspring. J Nutr 135:1382–1386

Lisle SJ, Lewis RM, Petry CJ, Ozanne SE, Hales CN, Forhead AJ (2003) Effect of maternal iron restriction during pregnancy on renal morphology in the adult rat offspring. Br J Nutr 90:33–39

Liu H, Wintour EM (2005) Aquaporins in development–a review. Reprod Biol Endocrinol 3:18

Liu J, Zhang L, Wang D, Shen H, Jiang M, Mei P, Hayden PS, Sedor JR, Hu H (2003) Congenital diaphragmatic hernia, kidney agenesis and cardiac defects associated with Slit3-deficiency in mice. Mech Dev 120:1059–1070

Liu L, Barajas L (1993) The rat renal nerves during development. Anat Embryol 188:345–361

Liu ZZ, Wada J, Kumar A, Carone FA, Takahashi M, Kanwar YS (1996) Comparative role of phosphotyrosine kinase domains of c-ros and c-Ret protooncogenes in metanephric development with respect to growth factors and matrix morphogens. Dev Biol 178:133–148

Long JZ, Lackan CS, Hadjantonakis AK (2005) Genetic and spectrally distinct in vivo imaging: embryonic stem cells and mice with widespread expression of a monomeric red fluorescent protein. BMC Biotechnol 5:20

Loughna S, Hardman P, Landels E, Jussila L, Alitalo K, Woolf AS (1997) A molecular and genetic analysis of renal glomerular capillary development. Angiogenesis 1:84–101

Loughna S, Landels E, Woolf AS (1996) Growth factor control of developing kidney endothelial cells. Exp Nephrol 4:112–118

Loughna S, Yuan HT, Woolf AS (1998) Effects of oxygen on vascular patterning in Tie1/LacZ metanephric kidneys in vitro. Biochem Biophys Res Commun 247:361–366

Lucas JS, Inskip HM, Godfrey KM, Foreman CT, Warner JO, Gregson RK, Clough JB (2004) Small size at birth and greater postnatal weight gain: relationships to diminished infant lung function. Am J Respir Crit Care Med 170:534–540

Ludwig KS, Landmann L (2005) Early development of the human mesonephros. Anat Embryol 209:439–447

Lumbers ER (1995) Functions of the renin–angiotensin system during development. Clin Exp Pharmacol Physiol 22:499–505

Maeshima A, Vaughn DA, Choi Y, Nigam SK (2006) Activin A is an endogenous inhibitor of ureteric bud outgrowth from the Wolffian duct. Dev Biol 295:473–485

Mah SP, Saueressig H, Goulding M, Kintner C, Dressler GR (2000) Kidney development in cadherin-6 mutants: delayed mesenchyme-to-epithelial conversion and loss of nephrons. Dev Biol 223:38–53

Mahoney ZX, Sammut B, Xavier RJ, Cunningham J, Go G, Brim KL, Stappenbeck TS, Miner JH, Swat W (2006) Discs-large homolog 1 regulates smooth muscle orientation in the mouse ureter. Proc Natl Acad Sci U S A 103:19872–19877

Majumdar A, Vainio S, Kispert A, McMahon J, McMahon AP (2003) Wnt11 and Ret/Gdnf pathways cooperate in regulating ureteric branching during metanephric kidney development. Development 130:3175–3185

Manning J, Beutler K, Knepper MA, Vehaskari VM (2002) Upregulation of renal BSC1 and TSC in prenatally programmed hypertension. Am J Physiol Renal Physiol 283: F202–F206

Maric C, Aldred GP, Harris PJ, Alcorn D (1998) Angiotensin II inhibits growth of cultured embryonic renomedullary interstitial cells through the AT2 receptor. Kidney Int 53:92–99

Martinez G, Cullen-McEwen L, Bertram JF (2001) Transforming growth factor-beta superfamily members: roles in branching morphogenesis in the kidney. Nephrology 6:274–284

Martinez G, Georgas K, Challen GA, Rumballe B, Davis MJ, Taylor D, Teasdale RD, Grimmond SM, Little MH (2006) Definition and spatial annotation of the dynamic secretome during early kidney development. Dev Dyn 235:1709–1719

Martyn CN, Barker DJ, Osmond C (1996) Mothers' pelvic size, fetal growth, and death from stroke and coronary heart disease in men in the UK. Lancet 348:1264–1268

Masuki S, Todo T, Nakano Y, Okamura H, Nose H (2005) Reduced alpha-adrenoceptor responsiveness and enhanced baroreflex sensitivity in Cry-deficient mice lacking a biological clock. J Physiol 566:213–224

Matsubara M (2004) Renal sodium handling for body fluid maintenance and blood pressure regulation. Yakugaku Zasshi 124:301–309

McConnell J (2006) A mitochondrial component of developmental programming. In: Gluckman P, Hanson M (eds) Developmental origins of health and disease. Cambridge University Press, Cambridge, pp 75–81

McCright B, Gao X, Shen L, Lozier J, Lan Y, Maguire M, Herzlinger D, Weinmaster G, Jiang R, Gridley T (2001) Defects in development of the kidney, heart and eye vasculature in mice homozygous for a hypomorphic Notch2 mutation. Development 128:491–502

McMullen S, Langley-Evans SC (2005) Sex-specific effects of prenatal low-protein and carbenoxolone exposure on renal angiotensin receptor expression in rats. Hypertension 46:1374–1380

Meaney MJ, Szyf M, Seckl JR (2007) Epigenetic mechanisms of perinatal programming of hypothalamic-pituitary-adrenal function and health. Trends Mol Med 13:269–277

Mendelsohn C, Batourina E, Fung S, Gilbert T, Dodd J (1999) Stromal cells mediate retinoid-dependent functions essential for renal development. Development 126:1139–1148

Menini S, Ricci C, Iacobini C, Bianchi G, Pugliese G, Pesce C (2004) Glomerular number and size in Milan hypertensive and normotensive rats: their relationship to susceptibility and resistance to hypertension and renal disease. J Hypertens 22:2185–2192

Merlet-Benichou C (1999) Influence of fetal environment on kidney development. Int J Dev Biol 43:453–456

Michos O, Goncalves A, Lopez-Rios J, Tiecke E, Naillat F, Beier K, Galli A, Vainio S, Zeller R (2007) Reduction of BMP4 activity by gremlin 1 enables ureteric bud outgrowth and GDNF/WNT11 feedback signalling during kidney branching morphogenesis. Development 134:2397–2405

Michos O, Panman L, Vintersten K, Beier K, Zeller R, Zuniga A (2004) Gremlin-mediated BMP antagonism induces the epithelial-mesenchymal feedback signaling controlling metanephric kidney and limb organogenesis. Development 131:3401–3410

Miner JH, Li C (2000) Defective glomerulogenesis in the absence of laminin alpha5 demonstrates a developmental role for the kidney glomerular basement membrane. Dev Biol 217:278–289

Mitchell EK, Louey S, Cock ML, Harding R, Black MJ (2004) Nephron endowment and filtration surface area in the kidney after growth restriction of fetal sheep. Pediatr Res 55:769–773

Miyamoto N, Yoshida M, Kuratani S, Matsuo I, Aizawa S (1997) Defects of urogenital development in mice lacking Emx2. Development 124:1653–1664

Miyazaki Y, Oshima K, Fogo A, Hogan BL, Ichikawa I (2000) Bone morphogenetic protein 4 regulates the budding site and elongation of the mouse ureter. J Clin Invest 105:863–873

Miyazaki Y, Oshima K, Fogo A, Ichikawa I (2003) Evidence that bone morphogenetic protein 4 has multiple biological functions during kidney and urinary tract development. Kidney Int 63:835–844

Miyazaki Y, Tsuchida S, Nishimura H, Pope JC, Harris RC, McKanna JM, Inagami T, Hogan BL, Fogo A, Ichikawa I (1998) Angiotensin induces the urinary peristaltic machinery during the perinatal period. J Clin Invest 102:1489–1497

Modi N (2003) Clinical implications of postnatal alterations in body water distribution. Semin Neonatol 8:301–306

Moore MW, Klein RD, Farinas I, Sauer H, Armanini M, Phillips H, Reichardt LF, Ryan AM, Carver-Moore K, Rosenthal A (1996) Renal and neuronal abnormalities in mice lacking GDNF. Nature 382:76–79

Moritz K, Koukoulas I, Albiston A, Wintour EM (2000) Angiotensin II infusion to the midgestation ovine fetus: effects on the fetal kidney. Am J Physiol Regul Integr Comp Physiol 279:R1290–R1297

Moritz KM, Bertram JF (2006) Barker and Brenner: a basis for hypertension? Curr Hypertens Rev 2:179–185

Moritz KM, Wintour EM (1999) Functional development of the meso- and metanephros. Pediatr Nephrol 13:171–178

Moritz KM, Boon WM, Wintour EM (2005a) Glucocorticoid programming of adult disease. Cell Tissue Res 322:81–88

Moritz KM, Dodic M, Wintour EM (2003) Kidney development and the fetal programming of adult disease. Bioessays 25:212–220

Moritz KM, Jefferies A, Wong J, Wintour EM, Dodic M (2005b) Reduced renal reserve and increased cardiac output in female sheep uninephrectomised as fetuses. Kidney Int 67:822–828

Moritz KM, Johnson K, Douglas-Denton R, Wintour EM, Dodic M (2002a) Maternal glucocorticoid treatment programs alterations in the renin-angiotensin system of the ovine fetal kidney. Endocrinology 143:4455–4463

Moritz KM, Wintour EM, Dodic M (2002b) Fetal uninephrectomy leads to postnatal hypertension and compromised renal function. Hypertension 39:1071–1076

Murotsuki J, Challis JR, Han VK, Fraher LJ, Gagnon R (1997) Chronic fetal placental embolization and hypoxemia cause hypertension and myocardial hypertrophy in fetal sheep. Am J Physiol 272:R201–R207

Murphy VE, Fittock RJ, Zarzycki PK, Delahunty MM, Smith R, Clifton VL (2007) Metabolism of synthetic steroids by the human placenta. Placenta 28:39–46

Naim E, Bernstein A, Bertram JF, Caruana G (2005) Mutagenesis of the epithelial polarity gene, discs large 1, perturbs nephrogenesis in the developing mouse kidney. Kidney Int 68:955–965. Erratum in: Kidney Int 68:1922

Nenov VD, Taal MW, Sakharova OV, Brenner BM (2000) Multi-hit nature of chronic renal disease. Curr Opin Nephrol Hypertens 9:85–97

Niimura F, Labosky PA, Kakuchi J, Okubo S, Yoshida H, Oikawa T, Ichiki T, Naftilan AJ, Fogo A, Inagami T, Hogan BLM, Ichikawa I (1995) Gene targeting in mice reveals a requirement for angiotensin in the development and maintenance of kidney morphology and growth factor regulation. J Clin Invest 96:2947–2954

Nishimura H, Yerkes E, Hohenfellner K, Miyazaki Y, Ma J, Hunley TE, Yoshida H, Ichiki T, Threadgill D, Phillips JA 3rd, Hogan BM, Fogo A, Brock JW 3rd, Inagami T, Ichikawa I (1999) Role of the angiotensin type 2 receptor gene in congenital anomalies of the kidney and urinary tract, CAKUT, of mice and men. Mol Cell 3:1–10

Nishinakamura R, Takasato M (2005) Essential roles of Sall1 in kidney development. Kidney Int 68:1948–1950

Nishinakamura R, Matsumoto Y, Nakao K, Nakamura K, Sato A, Copeland NG, Gilbert DJ, Jenkins NA, Scully S, Lacey DL, Katsuki M, Asashima M, Yokota T (2001) Murine homolog of SALL1 is essential for ureteric bud invasion in kidney development. Development 128:3105–3115

Noble L (2003) Developments in neonatal technology continue to improve infant outcomes. Pediatr Ann 32:595–603

Noia G, Masini L, Caruso A, Perrelli L, Calisti A, Salvaggio E, Mancuso S (1989) Prenatal diagnosis of congenital uropathies. Fetal Ther 1989 4:40–51

Nyengaard JR, Bendtsen TF (1992) Glomerular number and size in relation to age, kidney weight, and body surface in normal man. Anat Rec 232:194–201

Ohuchi H, Hori Y, Yamasaki M, Harada H, Sekine K, Kato S, Itoh N (2000) FGF10 acts as a major ligand for FGF receptor 2 IIIb in mouse multi-organ development. Biochem Biophys Res Commun 277:643–649

Oliverio MI, Kim HS, Ito M, Le T, Audoly L, Best CF, Hiller S, Kluckman K, Maeda N, Smithies O, Coffman TM (1998) Reduced growth, abnormal kidney structure, and type 2 (AT2) angiotensin receptor-mediated blood pressure regulation in mice lacking both AT1A and AT1B receptors for angiotensin II. Proc Natl Acad Sci U S A 95:15496–15501

Omar SA, DeCristofaro JD, Agarwal BI, La Gamma EF (1999) Effects of prenatal steroids on water and sodium homeostasis in extremely low birth weight neonates. Pediatrics 104:482–488

Ortiz LA, Quan A, Zarzar F, Weinberg A, Baum M (2003) Prenatal dexamethasone programs hypertension and renal injury in the rat. Hypertension 41:328–334

Osathanondh V, Potter EL (2007) Development of the human kidney as shown by microdissection III. Formation and interrelationships of collecting tubules and nephrons. Arch Pathol 76:290–302

Osmond C, Barker DJ, Winter PD, Fall CH, Simmonds SJ (1993) Early growth and death from cardiovascular disease in women. BMJ 307:1519–1524

Ozanne SE, Constancia M (2007) Mechanisms of disease: the developmental origins of disease and the role of the epigenotype. Nat Clin Pract Endocrinol Metab 3:539–546

Page WV, Perlman S, Smith FG, Segar JL, Robillard JE (1992) Renal nerves modulate kidney renin gene expression during the transition from fetal to newborn life. Am J Physiol 262: R459–R463

Patari-Sampo A, Ihalmo P, Holthofer H (2006) Molecular basis of the glomerular filtration: nephrin and the emerging protein complex at the podocyte slit diaphragm. Ann Med 38:483–492

Patterson LT, Pembaur M, Potter SS (2001) Hoxa11 and Hoxd11 regulate branching morphogenesis of the ureteric bud in the developing kidney. Development 128:2153–2161

Peers A, Hantzis V, Dodic M, Koukoulas I, Gibson A, Baird R, Salemi R, Wintour EM (2001) Functional glucocorticoid receptors in the mesonephros of the ovine fetus. Kidney Int 59:425–433

Perantoni AO, Dove LF, Karavanova I (1995) Basic fibroblast growth factor can mediate the early inductive events in renal development. Proc Natl Acad Sci U S A 92:4696–4700

Peters CA (2001) Animal models of fetal renal disease. Prenat Diagn 21:917–923

Pichel JG, Shen L, Sheng HZ, Granholm AC, Drago J, Grinberg A, Lee EJ, Huang SP, Saarma M, Hoffer BJ, Sariola H, Westphal H (1996) Defects in enteric innervation and kidney development in mice lacking GDNF. Nature 382:73–76

Piper M, Georgas K, Yamada T, Little M (2000) Expression of the vertebrate Slit gene family and their putative receptors, the Robo genes, in the developing murine kidney. Mech Dev 94:213–217

Piscione TD, Wu MY, Quaggin SE (2004) Expression of Hairy/Enhancer of Split genes, Hes1 and Hes5, during murine nephron morphogenesis. Gene Expr Patterns 4:707–711

Plank C, Ostreicher I, Hartner A, Marek I, Struwe FG, Amann K, Hilgers KF, Rascher W, Dotsch J (2006) Intrauterine growth retardation aggravates the course of acute mesangioproliferative glomerulonephritis in the rat. Kidney Int 70:1974–1982

Plisov SY, Yoshino K, Dove LF, Higinbotham KG, Rubin JS, Perantoni AO (2001) TGF beta 2, LIF and FGF2 cooperate to induce nephrogenesis. Development 128:1045–1057

Pohl M, Stuart RO, Sakurai H, Nigam SK (2000) Branching morphogenesis during kidney development. Annu Rev Physiol 62:595–620

Pole RJ, Qi BQ, Beasley SW (2002) Patterns of apoptosis during degeneration of the pronephros and mesonephros. J Urol 167:269–271

Pope JC 4th, Brock JW 3rd, Adams MC, Stephens FD, Ichikawa I (1999) How they begin and how they end: classic and new theories for the development and deterioration of congenital anomalies of the kidney and urinary tract, CAKUT. J Am Soc Nephrol 9:2018–2028

Pope JC 4th, Nishimura H, Ichikawa I (1998) Role of angiotensin in the development of the kidney and urinary tract. Nephrologie 19:433–436

Potter EL (1965) Bilateral absence of ureters and kidneys. Obstet Gynecol 25:3–12

Pugh JL, Sweeney WE Jr, Avner ED (1995) Tyrosine kinase activity of the EGF receptor in murine metanephric organ culture. Kidney Int 47:774–781

Qiao J, Sakurai H, Nigam SK (1999a) Branching morphogenesis independent of mesenchymal-epithelial contact in the developing kidney. Proc Natl Acad Sci U S A 96:7330–7335

Qiao J, Uzzo R, Obara-Ishihara T, Degenstein L, Fuchs E, Herzlinger D (1999b) FGF-7 modulates ureteric bud growth and nephron number in the developing kidney. Development 126:547–554

Quaggin SE, Schwartz L, Cui S, Igarashi P, Deimling J, Post M, Rossant J (1999) The basic-helix-loop-helix protein pod1 is critically important for kidney and lung organogenesis. Development 126:5771–5783

Quaggin SE, Yeger H, Igarashi P (1997) Antisense oligonucleotides to Cux-1, a Cut-related homeobox gene, cause increased apoptosis in mouse embryonic kidney cultures. J Clin Invest 99:718–724

Raatikainen-Ahokas A, Hytonen M, Tenhunen A, Sainio K, Sariola H (2000) BMP-4 affects the differentiation of metanephric mesenchyme and reveals an early anterior-posterior axis of the embryonic kidney. Dev Dyn 217:146–158

Rascle A, Suleiman H, Neumann T, Witzgall R (2007) Role of transcription factors in podocytes. Nephron Exp Nephrol 106:e60–e66

Revest JM, Spencer-Dene B, Kerr K, De Moerlooze L, Rosewell I, Dickson C (2001) Fibroblast growth factor receptor 2-IIIb acts upstream of Shh and Fgf4 and is required for limb bud maintenance but not for the induction of Fgf8, Fgf10, Msx1, or Bmp4. Dev Biol 231:47–62

Robillard JE, Guillery EN, Segar JL, Merrill DC, Jose PA (1993) Influence of renal nerves on renal function during development. Pediatr Nephrol 7:667–671

Robinson JS, Kingston EJ, Jones CT, Thorburn GD (1979) Studies on experimental growth retardation in sheep. The effect of removal of endometrial caruncles on fetal size and metabolism. J Dev Physiol 1:379–398

Rodriguez MM, Gomez A, Abitbol C, Chandar J, Montane B, Zilleruelo G (2005) Comparative renal histomorphometry: a case study of oligonephropathy of prematurity. Pediatr Nephrol 20:945–949

Rousham EK, Gracey M (2002) Factors affecting birthweight of rural Australian Aborigines. Ann Human Biol 29:363–372

Roy S 3rd, Pitcock JA, Etteldorf JN (1976) Prognosis of acute poststreptococcal glomerulonephritis in childhood: prospective study and review of the literature. Adv Pediatr 23:35–69

Sahajpal V, Ashton N (2003) Renal function and angiotensin AT1 receptor expression in young rats following intrauterine exposure to a maternal low-protein diet. Clin Sci 104:607–614

Saifur Rohman M, Emoto N, Nonaka H, Okura R, Nishimura M, Yagita K, van der Horst GT, Matsuo M, Okamura H, Yokoyama M (2005) Circadian clock genes directly regulate expression of the Na(+)/H(+) exchanger NHE3 in the kidney. Kidney Int 67:1410–1419

Sainio K (2003) Development of the mesonephric kidney. In: Vize PD, Woolf AS, Bard JBL (eds) The kidney: from normal development to congenital disease. Academic Press, London, pp 75–86

Sainio K, Raatikainen-Ahokas A (1999) Mesonephric kidney–a stem cell factory? Int J Dev Biol 43:435–439

Sainio K, Saarma M, Nonclercq D, Paulin L, Sariola H (1994) Antisense inhibition of low-affinity nerve growth factor receptor in kidney cultures: power and pitfalls. Cell Mol Neurobiol 14:439–457

Sainio K, Suvanto P, Davies J, Wartiovaara J, Wartiovaara K, Saarma M, Arumae U, Meng X, Lindahl M, Pachnis V, Sariola H (1997) Glial-cell-line-derived neurotrophic factor is required for bud initiation from ureteric epithelium. Development 124:4077–4087

Sajithlal G, Zou D, Silvius D, Xu PX (2005) Eya 1 acts as a critical regulator for specifying the metanephric mesenchyme. Dev Biol 284:323–336

Samuel T, Hoy WE, Douglas-Denton R, Hughson MD, Bertram JF (2005) Determinants of glomerular volume in different cortical zones of the human kidney. J Am Soc Nephrol 16:3102–3109

Sanchez MP, Silos-Santiago I, Frisen J, He B, Lira SA, Barbacid M (1996) Renal agenesis and the absence of enteric neurons in mice lacking GDNF. Nature 382:70–73

Sanford LP, Ormsby I, Gittenberger-de Groot AC, Sariola H, Friedman R, Boivin GP, Cardell EL, Doetschman T (1997) TGFbeta2 knockout mice have multiple developmental defects that are non-overlapping with other TGFbeta knockout phenotypes. Development 124:2659–2670

Sanna-Cherchi S, Caridi G, Weng PL, Scolari F, Perfumo F, Gharavi AG, Ghiggeri GM (2007) Genetic approaches to human renal agenesis/hypoplasia and dysplasia. Pediatr Nephrol 22:1675–1684

Sariola H (2002) Nephron induction revisited: from caps to condensates. Curr Opin Nephrol Hypertens 11:17–21

Sariola H, Saarma M (1999) GDNF and its receptors in the regulation of the ureteric branching. Int J Dev Biol 43:413–418

Sariola H, Aufderheide E, Bernhard H, Henke-Fahle S, Dippold W, Ekblom P (1988) Antibodies to cell surface ganglioside GD3 perturb inductive epithelial-mesenchymal interactions. Cell 54:235–245

Sariola H, Ekblom P, Lehtonen E, Saxén L (1983) Differentiation and vascularization of the metanephric kidney grafted on the chorioallantoic membrane. Dev Biol 96:427–435

Sariola H, Saarma M, Sainio K, Arumae U, Palgi J, Vaahtokari A, Thesleff I, Karavanov A (1991) Dependence of kidney morphogenesis on the expression of nerve growth factor receptor. Science 254:571–573

Sawdy RJ, Lye S, Fisk NM, Bennett PR (2003) A double-blind randomized study of fetal side effects during and after the short-term maternal administration of indomethacin, sulindac, and nimesulide for the treatment of preterm labor. Am J Obstet Gynecol 188:1046–1051

Saxén L, Sariola H (1987) Early organogenesis of the kidney. Pediatr Nephrol 1:385–392

Schmidt-Ott KM, Lan D, Hirsh BJ, Barasch J (2006) Dissecting stages of mesenchymal-to-epithelial conversion during kidney development. Nephron Physiol 104:56–60

Schmidt-Ott KM, Yang J, Chen X, Wang H, Paragas N, Mori K, Li JY, Lu B, Costantini F, Schiffer M, Bottinger E, Barasch J (2005) Novel regulators of kidney development from the tips of the ureteric bud. J Am Soc Nephrol 16:1993–2002

Schnabel CA, Godin RE, Cleary ML (2003) Pbx1 regulates nephrogenesis and ureteric branching in the developing kidney. Dev Biol 254:262–276

Schreuder MF, Nyengaard JR, Fodor M, van Wijk JA, Delemarre-van de Waal HA (2005) Glomerular number and function are influenced by spontaneous and induced low birth weight in rats. J Am Soc Nephrol 16:2913–2919

Schreuder MF, Nyengaard JR, Remmers F, van Wijk JA, Delemarre-van de Waal HA (2006) Postnatal food restriction in the rat as a model for a low nephron endowment. Am J Physiol Renal Physiol 291:F1104–F1107

Schuchardt A, D'Agati V, Pachnis V, Costantini F (1996) Renal agenesis and hypodysplasia in ret-k- mutant mice result from defects in ureteric bud development. Development 122:1919–1929

Schutz S, Le Moullec J-M, Corvol P, Gasc J-M (1996) Early expression of all the components of the renin–angiotensin system in human development. Am J Pathol 149:2067–2079

Schwab K, Hartman HA, Liang HC, Aronow BJ, Patterson LT, Potter SS (2006) Comprehensive microarray analysis of Hoxa11/Hoxd11 mutant kidney development. Dev Biol 293:540–554

Schwab K, Patterson LT, Aronow BJ, Luckas R, Liang HC, Potter SS (2003) A catalogue of gene expression in the developing kidney. Kidney Int 64:1588–1604

Self M, Lagutin OV, Bowling B, Hendrix J, Cai Y, Dressler GR, Oliver G (2006) Six2 is required for suppression of nephrogenesis and progenitor renewal in the developing kidney. EMBO J 25:5214–5228

Shakya R, Watanabe T, Costantini F (2005) The role of GDNF/Ret signalling in ureteric bud cell fate and branching morphogenesis. Dev Cell 8:65–74

Shao X, Johnson JE, Richardson JA, Hiesberger T, Igarashi P (2002) A minimal Ksp-cadherin promoter linked to a green fluorescent protein reporter gene exhibits tissue-specific expression in the developing kidney and genitourinary tract. J Am Soc Nephrol 13:1824–1836

Shawlot W, Behringer RR (1995) Requirement for Lim1 in head-organizer function. Nature 374:425–430

Shotan A, Widerhorn J, Hurst A, Elkayam U (1994) Risks of angiotensin-converting enzyme inhibition during pregnancy: experimental and clinical evidence, potential mechanisms, and recommendations for use. Am J Med 96:451–456

Singh RR, Cullen-McEwen LA, Kett MM, Boon WM, Dowling J, Bertram JF, Moritz KM (2007a) Prenatal corticosterone exposure results in altered AT1/AT2, nephron deficit and hypertension in the rat offspring. J Physiol 579:503–513

Singh RR, Moritz KM, Bertram JF, Cullen-McEwen LA (2007b) Effects of dexamethasone exposure on rat metanephric development: in vitro and in vivo studies. Am J Physiol Renal Physiol 293:F548–F554

Smith FG, Sato T, McWeeny OJ, Klinkefus JM, Robillard JE (1990) Role of renal sympathetic nerves in response of the ovine fetus to volume expansion. Am J Physiol 259:R1050–R1055

Smith RM, Smith PA, McKinnon M, Gracey M (2000) Birthweights and growth of infants in five Aboriginal communities. Aust N Z J Public Health 24:124–135

Smyth I, Du X, Taylor MS, Justice MJ, Beutler B, Jackson IJ (2004) The extracellular matrix gene Frem1 is essential for the normal adhesion of the embryonic epidermis. Proc Natl Acad Sci U S A 101:13560–13565

Song J, Oh JY, Sung YA, Pak YK, Park KS, Lee HK (2001) Peripheral blood mitochondrial DNA content is related to insulin sensitivity in offspring of type 2 diabetic patients. Diabetes Care 24:865–869

Sorokin L, Sonnenberg A, Aumailley M, Timpl R, Ekblom P (1990) Recognition of the laminin E8 cell-binding site by an integrin possessing the alpha 6 subunit is essential for epithelial polarization in developing kidney tubules. J Cell Biol 111:1265–1273

Srinivas S, Goldberg MR, Watanabe T, D'Agati V, al-Awqati Q, Costantini F (1999a) Expression of green fluorescent protein in the ureteric bud of transgenic mice: a new tool for the analysis of ureteric bud morphogenesis. Dev Genet 24:241–251

Srinivas S, Wu Z, Chen CM, D'Agati V, Costantini F (1999b) Dominant effects of RET receptor misexpression and ligand-independent RET signaling on ureteric bud development. Development 126:1375–1386

Stark K, Vainio S, Vassileva G, McMahon AP (1994) Epithelial transformation of metanephric mesenchyme in the developing kidney regulated by Wnt-4. Nature 372:679–683

Sterio DC (1984) The unbiased estimation of number and sizes of arbitrary particles using the disector. J Microsc 134:127–136

Strube YN, Beard JL, Ross AC (2002) Iron deficiency and marginal vitamin A deficiency affect growth, hematological indices and the regulation of iron metabolism genes in rats. J Nutr 132:3607–3615

Stuart RO, Bush KT, Nigam SK (2003) Changes in gene expression patterns in the ureteric bud and metanephric mesenchyme in models of kidney development. Kidney Int 64:1997–2008

Sumová A, Bendová Z, Sládek M, Kováciková Z, El-Hennamy R, Laurinová K, Illnerová H (2006) The rat circadian network and its photoperiodic entrainment during development. Chronobiol Int 23:237–243

Susser E, Neugebauer R, Hoek HW, Brown AS, Lin S, Labovitz D, Gorman JM (1996) Schizophrenia after prenatal famine. Further evidence. Arch Gen Psychiatry 53:25–31

Takamiya K, Kostourou V, Adams S, Jadeja S, Chalepakis G, Scambler PJ, Huganir RL, Adams RH (2004) A direct functional link between the multi-PDZ domain protein GRIP1 and the Fraser syndrome protein Fras1. Nat Genet 36:172–177

Takasato M, Osafune K, Matsumoto Y, Kataoka Y, Yoshida N, Meguro H, Aburatani H, Asashima M, Nishinakamura R (2004) Identification of kidney mesenchymal genes by a combination of microarray analysis and Sall1-GFP knockin mice. Mech Dev 121:547–557

Takemoto M, He L, Norlin J, Patrakka J, Xiao Z, Petrova T, Bondjers C, Asp J, Wallgard E, Sun Y, Samuelsson T, Mostad P, Lundin S, Miura N, Sado Y, Alitalo K, Quaggin SE, Tryggvason K, Betsholtz C (2006) Large-scale identification of genes implicated in kidney glomerulus development and function. EMBO J 25:1160–1174

Tarry-Adkins JL, Joles JA, Chen JH, Martin-Gronert MS, van der Giezen DM, Goldschmeding R, Hales CN, Ozanne SE (2007) Protein restriction in lactation confers nephroprotective effects in the male rat and is associated with increased antioxidant expression. Am J Physiol Regul Integr Comp Physiol 293:R1259–R1266

Taylor PD, McConnell J, Khan IY, Holemans K, Lawrence KM, Asare-Anane H, Persaud SJ, Jones PM, Petrie L, Hanson MA, Poston L (2005) Impaired glucose homeostasis and mitochondrial abnormalities in offspring of rats fed a fat-rich diet in pregnancy. Am J Physiol Regul Integr Comp Physiol 288:R134–R139

Tiniakos D, Anagnostou V, Stavrakis S, Karandrea D, Agapitos E, Kittas C (2004) Ontogeny of intrinsic innervation in the human kidney. Anat Embryol 209:41–47

Tonkiss J, Trzci ska M, Galler JR, Ruiz-Opazo N, Herrera VL (1998) Prenatal malnutrition-induced changes in blood pressure: dissociation of stress and nonstress responses using radiotelemetry. Hypertension 32:108–114

Torres M, Gómez-Pardo E, Dressler GR, Gruss P (1995) Pax-2 controls multiple steps of urogenital development. Development 121:4057–4065

Torres-Farfan C, Rocco V, Monso C, Valenzuela FJ, Campino C, Germain A, Torrealba F, Valenzuela GJ, Seron-Ferre M (2006) Maternal melatonin effects on clock gene expression in a non-human primate fetus. Endocrinology 147:4618–4626

Towstoless MK, McDougall JG, Wintour EM (1989) Gestational changes in renal responsiveness to cortisol in the ovine fetus. Pediatr Res 26:6–10

Trifunovic A, Wredenberg A, Falkenberg M, Spelbrink JN, Rovio AT, Bruder CE, Bohlooly YM, Gidlof S, Oldfors A, Wibom R, Tornell J, Jacobs HT, Larsson NG (2004) Premature ageing in mice expressing defective mitochondrial DNA polymerase. Nature 429:417–423

Tsang TE, Shawlot W, Kinder SJ, Kobayashi A, Kwan KM, Schughart K, Kania A, Jessell TM, Behringer RR, Tam PP (2000) Lim1 activity is required for intermediate mesoderm differentiation in the mouse embryo. Dev Biol 223:77–90

Tufro-McReddie A, Romano LM, Harris JM, Ferder L, Gomez RA (1995) Angiotensin II regulates nephrogenesis and renal vascular development. Am J Physiol Renal Physiol 269:F110–F115

Valerius MT, Patterson LT, Witte DP, Potter SS (2002) Microarray analysis of novel cell lines representing two stages of metanephric mesenchyme differentiation. Mech Dev 112:219–232

van den Broek N (2003) Anaemia and micronutrient deficiencies. Br Med Bull 67:149–160

Vehaskari VM, Stewart T, Lafont D, Soyez C, Seth D, Manning J (2004) Kidney angiotensin and angiotensin receptor expression in prenatally programmed hypertension. Am J Physiol Renal Physiol 287:F262–F267

Vickaryous V, Whitelaw E (2006) Modification of epigenetic state through dietary manipulation in the developing mammalian embryo. In: Wintour EM, Owens JA (eds) Early life origins of health and disease. Landes Bioscience, New York, pp 70–78

Vilar J, Gilbert T, Moreau E, Merlet-Benichou C (1996) Metanephros organogenesis is highly stimulated by vitamin A derivatives in organ culture. Kidney Int 49:1478–1487

Vilar J, Lalou C, Duong VH, Charrin S, Hardouin S, Raulais D, Merlet-Bénichou C, Lelièvre-Pégorier M (2002) Midkine is involved in kidney development and in its regulation by retinoids. J Am Soc Nephrol 13:668–676

Vize PD, Carrol TJ, Wallingford JB (2003) Induction, development, and physiology of the pronephric tubules. In: Vize PD, Woolf AS, Bard JBL (eds) The kidney: from normal development to congenital disease. Academic Press, London, pp 19–50

Vize PD, Seufert DW, Carroll TJ, Wallingford JB (1997) Model systems for the study of kidney development: use of the pronephros in the analysis of organ induction and patterning. Dev Biol 188:189–204

Walsh TJ, Hsieh S, Grady R, Mueller BA (2007) Antenatal hydronephrosis and the risk of pyelonephritis hospitalization during the first year of life. Urology 69:970–974

Wang Q, Lan Y, Cho ES, Maltby KM, Jiang R (2005) Odd-skipped related 1 (Odd 1) is an essential regulator of heart and urogenital development. Dev Biol 288:582–594

Watanabe T, Costantini F (2004) Real-time analysis of ureteric bud branching morphogenesis in vitro. Dev Biol 27:98–108

Waterland RA (2006) Critical experiments to determine if early nutritional influences on epigenetic mechanisms cause metabolic imprinting in humans. In: Wintour EM, Owens JA (eds) Early life origins of health and disease. Landes Bioscience, New York, pp 79–86

Watnick T, Germino GG (1999) Molecular basis of autosomal dominant polycystic kidney disease. Semin Nephrol. 19:327–343

Weaver IC, Cervoni N, Champagne FA, D'Alessio AC, Sharma S, Seckl JR, Dymov S, Szyf M, Meaney MJ (2004) Epigenetic programming by maternal behavior. Nat Neurosci 7:847–854

Weaver IC, Champagne FA, Brown SE, Dymov S, Sharma S, Meaney MJ, Szyf M (2005) Reversal of maternal programming of stress responses in adult offspring through methyl supplementation: altering epigenetic marking later in life. J Neurosci 25:11045–11054

Weaver IC, D'Alessio AC, Brown SE, Hellstrom IC, Dymov S, Sharma S, Szyf M, Meaney MJ (2007) The transcription factor nerve growth factor-inducible protein a mediates epigenetic programming: altering epigenetic marks by immediate-early genes. J Neurosci 27:1756–1768

Wei LN (2004) Retinoids and receptor interacting protein 140 (RIP140) in gene regulation. Curr Med Chem 11:1527–1532

Weibel ER (1979) Stereological methods, vol 1. Practical methods for biological morphometry. Academic Press, London.

Weibel ER, Gomez DM (1962) A principle for counting tissue structures on random sections. J Appl Physiol 17:343–348

Welham S, Riley PR, Wade A, Hubank M, Woolf AS (2005) Maternal diet programs embryonic kidney gene expression. Physiol Genomics 22:48–56

Welham SJ, Wade A, Woolf AS (2002) Protein restriction in pregnancy is associated with increased apoptosis of mesenchymal cells at the start of rat metanephrogenesis. Kidney Int 61:1231–1242

Wintour EM (1998) Water and electrolyte metabolism in the fetal-placental unit. In: Cowett RM (ed) Principles of perinatal, neonatal metabolism, 2nd edn. Springer-Verla, Berlin Heidelberg New York, pp 511–553

Wintour EM, Alcorn D, Rockell MD (1998) Development and function of the fetal kidney. In: Brace RA, Hanson MA, Rodeck CH (eds) Fetus and neonate, physiology and clinical applications, vol 4. Body fluids and kidney function. Cambridge University Press, Cambridge, pp 3–56

Wintour EM, Moritz KM, Johnson K, Ricardo S, Samuel CS, Dodic M (2003) Reduced nephron number in adult sheep, hypertensive as a result of prenatal glucocorticoid treatment. J Physiol 549:929–935

Wlodek ME, Mibus A, Tan A, Siebel AL, Owens JA, Moritz KM (2007) Normal lactational environment restores nephron endowment and prevents hypertension after placental restriction in the rat. J Am Soc Nephrol 18:1688–1696

Wlodek ME, Westcott KT, O'Dowd R, Serruto A, Wassef L, Moritz KM, Moseley JM (2005) Uteroplacental restriction in the rat impairs fetal growth in association with alterations in placental growth factors including PTHrP. Am J Physiol Regul Integr Comp Physiol 288:R1620–R1627

Wolf G, Neilson EG (1993) Angiotensin II as a renal growth factor. J Am Soc Nephrol 3:1531–1540

Woods LL, Ingelfinger JR, Rasch R (2005) Modest maternal protein restriction fails to program adult hypertension in female rats. Am J Physiol Regul Integr Comp Physiol 289: R1131–R1136

Woods LL, Weeks DA, Rasch R (2004) Programming of adult blood pressure by maternal protein restriction: role of nephrogenesis. Kidney Int 65:1339–1348

Woods LL, Weeks DA, Rasch R, Nyengaard JR, Rasch R (2001) Maternal protein restriction suppresses the newborn rennin-angiotensin system and programs adult hypertension in rats. Pediatr Res 49:460–467

Woolf AS (2001) The life of the human kidney before birth: its secrets unfold. Pediatr Res 49:8–10

Woolf AS, Price KL, Scambler PJ, Winyard PJ (2004) Links evolving concepts in human renal dysplasia. J Am Soc Nephrol 15:998–1007

Woolf AS (2006) Renal hypoplasia and dysplasia: starting to put the puzzle together. J Am Soc Nephrol 17:2647–2649

Woolf AS, Loughna S (1998) Origin of glomerular capillaries: is the verdict in? Exp Nephrol 6:17–21

Woolf AS, Winyard PJD, Hermans MM, Welham S (2003) Maldevelopment of the human kidney and lower urinary tract: an overview. In: Vize PD, Woolf AS, Bard JBL (eds) The kidney: from normal development to congenital disease. Academic Press, London, pp 377–393

Wrobel KH (2001) Morphogenesis of the bovine rete testis: extratesticular rete, mesonephros and establishment of the definitive urogenital junction. Anat Embryol 203:293–307

Wu Y-H, Fischer DF, Kalsbeek A, Garidou-Boof M-L, van der Vliet J, van Heijningen C, Liu R-Y, Zhou J-N, Swaab DF (2006) Pineal clock gene oscillation is disturbed in Alzheimer's disease, due to functional disconnection from the 'master clock'. FASEB J 20:1874–1876

Xu J, Desir GV (2007) Renalase, a new renal hormone: its role in health and disease. Curr Opin Nephrol Hypertens 16:373–378

Xu J, Li G, Wang P, Velazquez H, Yao X, Li Y, Wu Y, Peixoto A, Crowley S, Desir GV (2005) Renalase is a novel, soluble monoamine oxidase that regulates cardiac function and blood pressure. J Clin Invest 115:1275–1280

Xu PX, Adams J, Peters H, Brown MC, Heaney S, Maas R (1999) Eya1-deficient mice lack ears and kidneys and show abnormal apoptosis of organ primordia. Nat Genet 23:113–117

Xu PX, Zheng W, Huang L, Maire P, Laclef C, Silvius D (2003) Six1 is required for the early organogenesis of mammalian kidney. Development 130:3085–3094

Yosypiv IV, El-Dahr WE (2005) Role of the rennin-angiotensin system in the development of the uteric bud and renal collecting system. Pediatr Nephrol 20:1219–1229

Young RJ, Hoy WE, Kincaid-Smith P, Seymour AE, Bertram JF (2000) Glomerular size and glomerulosclerosis in Australian aborigines. Am J Kidney Dis 36:481–489

Yu J, Carroll TJ, McMahon AP (2002) Sonic hedgehog regulates proliferation and differentiation of mesenchymal cells in the mouse metanephric kidney. Development 129:5301–5312

Yuan HT, Suri C, Landon DN, Yancopoulos GD, Woolf AS (2000) Angiopoietin-2 is a site-specific factor in differentiation of mouse renal vasculature. J Am Soc Nephrol 11:1055–1066

Yuan HT, Suri C, Yancopoulos GD, Woolf AS (1999) Expression of angiopoietin-1, angiopoietin-2, and the Tie-2 receptor tyrosine kinase during mouse kidney maturation. J Am Soc Nephrol 10:1722–1736

Zandi-Nejad K, Luyckx VA, Brenner BM (2006) Adult hypertension and kidney disease: the role of fetal programming. Hypertension 47:502–508

Zhang H, Darwanto A, Linkhart TA, Sowers LC, Zhang L (2007) Maternal cocaine administration causes an epigenetic modification of protein kinase Cepsilon gene expression in fetal rat heart. Mol Pharmacol 71:1319–1328

Zhang SL, Guo J, Moini B, Ingelfinger JR (2004a) Angiotensin II stimulates Pax-2 in rat kidney proximal tubular cells: impact on proliferation and apoptosis. Kidney Int 66:2181–2192

Zhang SL, Moini B, Ingelfinger JR (2004b) Angiotensin II increases Pax-2 expression in fetal kidney cells via the AT2 receptor. J Am Soc Nephrol 15:1452–1465

Zhu X, Cheng J, Gao J, Lepor H, Zhang ZT, Pak J, Wu XR (2002) Isolation of mouse THP gene promoter and demonstration of its kidney-specific activity in transgenic mice. Am J Physiol Renal Physiol 282:F608–F617

Zimanyi MA, Bertram JF, Black MJ (2002) Nephron number and blood pressure in rat offspring with maternal high-protein diet. Pediatr Nephrol 17:1000–1004

Zimanyi MA, Bertram JF, Black MJ (2004) Does a nephron deficit predispose to salt-sensitive hypertension? Kidney Blood Press Res 27:239–247

Zimanyi MA, Denton KM, Forbes JM, Thallas-Bonke V, Thomas MC, Poon F, Black MJ (2006) A developmental nephron deficit in rats is associated with increased susceptibility to a secondary renal injury due to advanced glycation end-products. Diabetologia 49:801–810

Zohdi V, Moritz KM, Bubb KJ, Cock ML, Wreford N, Harding R, Black MJ (2007) Nephrogenesis and the renal renin-angiotensin system in fetal sheep: Effects of intrauterine growth restriction during late gestation. Am J Physiol Regul Integr Comp Physiol 293: R1267–R1273

Subject Index

MIX
Papier aus verantwortungsvollen Quellen
Paper from responsible sources
FSC® C105338

If you have any concerns about our products,
you can contact us on
ProductSafety@springernature.com

In case Publisher is established outside the EU,
the EU authorized representative is:
Springer Nature Customer Service Center GmbH
Europaplatz 3, 69115 Heidelberg, Germany

Printed by Libri Plureos GmbH
in Hamburg, Germany